FM 23-40

WAR DEPARTMENT FIELD MANUAL

THOMPSON SUBMACHINE GUN CALIBER .45 M1928A1

WAR DEPARTMENT · 31 DECEMBER 1941

DISCLAIMER:

THIS MANUAL IS SOLD FOR HISTORIC
RESEARCH PURPOSES ONLY, AS AN
ENTERTAINMENT. IT CONTAINS OBSOLETE
INFORMATION AND IS NOT INTENDED TO BE
USED AS PART OF AN ACTUAL OPERATION
OF MAINTENANCE TRAINING PROGRAM. NO
BOOK CAN SUBSTITUTE FOR PROPER
TRAINING BY AN AUTHORIZED INSTRUCTOR.

WAR DEPARTMENT FIELD MANUAL

FM 23-40

THOMPSON SUBMACHINE GUN CALIBER .45 M1928A1

WAR DEPARTMENT - 31 DECEMBER 1941

United States Government Printing Office
Washington : 1947

WAR DEPARTMENT,
WASHINGTON, December 31, 1941.

FM 23-40, Thompson Submachine Gun, Caliber .45, M1928A1, is published for the information and guidance of all concerned.

[A. G. 062.11 (8-12-41).]

BY ORDER OF THE SECRETARY OF WAR:

G. C. MARSHALL,
Chief of Staff.

OFFICIAL:

E. S. ADAMS,
Major General,
The Adjutant General.

DISTRIBUTION:

B1 (2), 2, 6, and 17 (5), 7 (3); R1 (2), 2 (5), 7 (3), 17 (10); IR 5 (5), 6 (10); Bn. 2, 9, and 17 (5), 7 (3); IBn 1, 5, 6, and 11 (5); C 9 (2), 17 (20); IC 2 (20), 5, 6, and 11 (10), 7 (2).

(For explanation of symbols see FM 21-6.)

TABLE OF CONTENTS

BASIC FIELD MANUAL

THOMPSON SUBMACHINE GUN, CALIBER .45, M1928A1

(This manual supersedes FM 23–40, August 19, 1940)

CHAPTER 1

MECHANICAL TRAINING

SECTION I

DESCRIPTION

■ 1. GENERAL.—The Thompson submachine gun, caliber .45, M1928A1 (fig. 1), is an air-cooled, recoil-operated, magazine-fed weapon weighing about 15¾ pounds with loaded 50-round magazine. The exterior surface of the rear portion of the barrel contains a series of annular flanges which serve to dissipate heat and cool the barrel during firing. The hand of the gunner is protected on the under side of the barrel by a wooden fore grip. A rear grip is also provided. Sling swivels are attached to the fore grip and stock for attaching the gun sling. The weapon is provided with a "Cutts" recoil compensator to lessen the recoil and the tendency of the muzzle to rise in full automatic fire. By correctly setting the rocker pivot, the weapon may be used for either full automatic or semiautomatic fire. The weapon is fed from a drum type magazine having a capacity of 50 rounds or from a box type magazine having a capacity of 20 rounds.

■ 2. GENERAL DATA.—*a. Dimension.*

(1) *Barrel.*

Diameter of bore....................inches.. 0.45

Number of grooves............................ 6

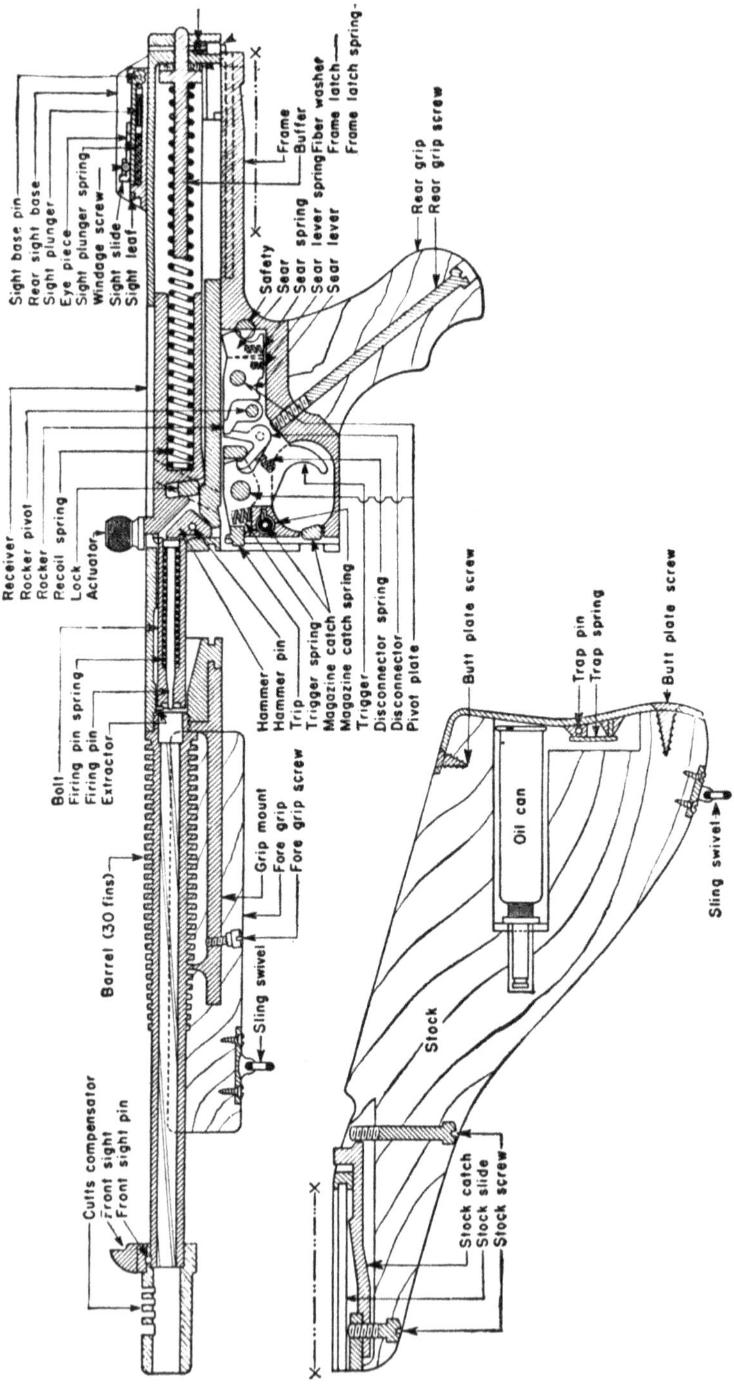

FIGURE 1.—Thompson submachine gun, caliber .45, M1928A1.

■ 2. General Data.—Continued.

(1) *Barrel*—Continued.

Twist in rifling, uniform, right, one turn in ....inches..   16

Length of barrel......................do.... 10.50

(2) *Gun.*

Over-all length of gun including compensator ....inches.. 33.69

Sight radius ....do.... 22.30

b. *Weight.*

Gun without magazine................pounds.. 10.75

Loaded 20-round magazine............do....   1.31

Loaded 50-round magazine............do....   4.95

Empty 20-round magazine.............do....   .38

Empty 50-round magazine.............do....   2.63

■ 3. Miscellaneous Data.

Initial velocity..............feet per second..   802

Pressure in chamber (approx.) pounds per square inch....................12,000-16,000

Weight of ball cartridge (approx.).....grains..   325

Weight of bullet (approx.)..............do....   234

Weight of powder charge (approx.)......do....   5

Rate of automatic fire (cyclic rate), shots per minute............................... 600–700

■ 4. Nomenclature of Component Parts.—*a. Gun* (figs. 1 and 2).

1. Actuator, assembly.
2. Barrel (fig. 1).
3. Bolt.
4. Breech oiler (with felt pads).
5. Buffer and fiber disk (the short end of the buffer is referred to as the buffer pilot, the long end as the buffer guide).
6. Disconnector.
7. Disconnector spring.
8. Ejector.
9. Extractor.
10. Firing pin.
11. Firing pin spring.
12. Fore grip (fig. 1).
13. Fore grip screw (fig. 1).
14. Frame (fig. 1).
15. Frame latch.
16. Frame latch spring.
17. Grip mount (fig. 1).
18. Hammer.
19. Hammer pin.
20. Lock.
21. Magazine catch, assembly
22. Magazine catch spring.
23. Pivot plate, assembly.
24. Rear grip (fig. 1).
25. Rear grip screw (fig. 1).

26. Receiver (fig. 1).
27. Recoil spring.
28. Rocker.
29. Rocker pivot or fire control lever.
30. Safety.
31. Sear.
32. Sear spring.
33. Sear lever.
34. Sear lever spring.
35. Trigger.
36. Trigger spring.
37. Trip.

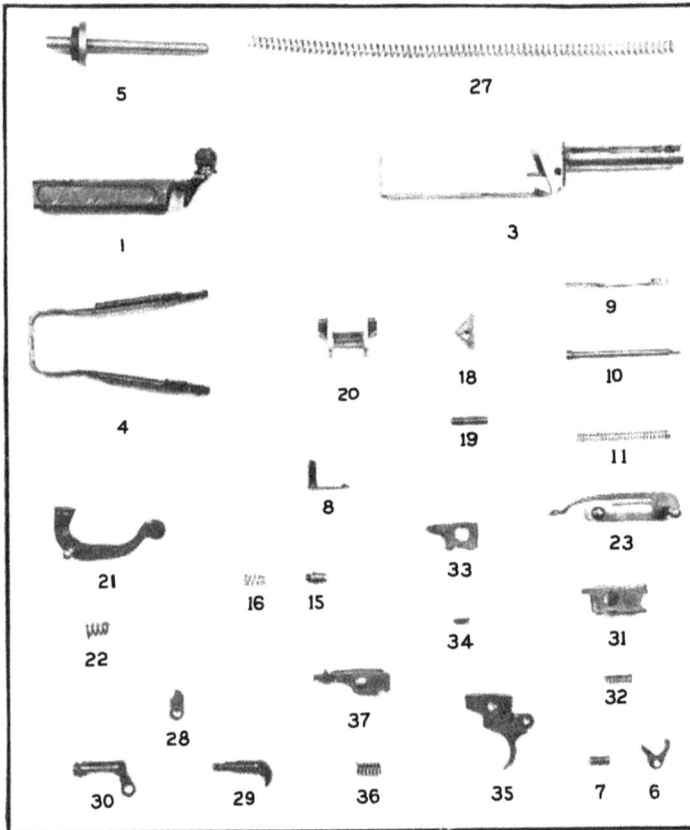

FIGURE 2.—Parts of the gun.

b. *Box magazine* (fig. 3).

38. Tube, assembly.
39. Follower.
40. Spring.
41. Floor plate.

c. *Drum magazine* (50 rounds) fig. 3).

42. Magazine body, assembly.
43. Cover, assembly.
44. Rotor, assembly.

45. Winding key, assembly.
46. Rotor retainer.

d. *Sights* (not separately illustrated).

47. Front sight (assembled to barrel) (fig. 1).
48. Rear sight (fig. 1).
49. Eyepiece.
50. Rear sight base (assembled to receiver).
51. Sight base pin.
52. Rear sight leaf plunger (with sight plunger pin).

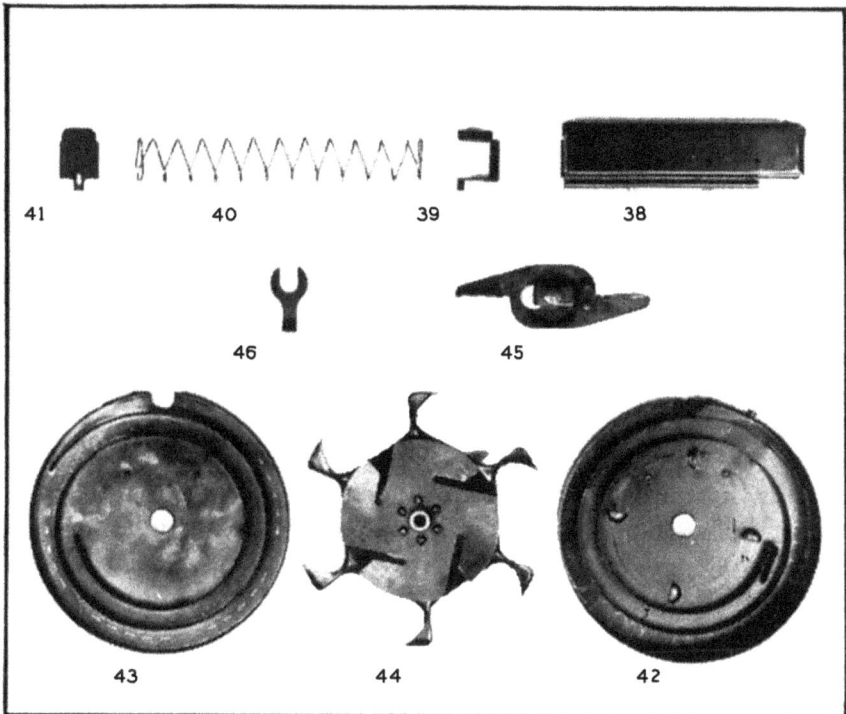

FIGURE 3.—Component parts of the magazines.

53. Sight plunger spring.
54. Sight leaf (with slide retaining pin).
55. Sight slide.
56. Sight slide catch.
57. Sight slide catch screw.
58. Windage screw (assembled), consisting of—
59. Windage screw.
60. Windage screw collar.
61. Windage screw collar pin.

SECTION II

DISASSEMBLING AND ASSEMBLING

■ 5. GENERAL.—*a.* Disassembling is considered under two general headings, removal of groups and disassembling of the groups.

b. A group is a number of components which either function together in the gun or are intimately related to each other and should therefore be considered together.

■ 6. REMOVAL OF GROUPS FROM GUN.—*a. Magazine.*—With the bolt in the rearward position, raise magazine catch and slide out magazine.

b. Stock.—Press in on butt stock catch button and slide butt stock to the rear.

c. Frame from receiver.—Turn safety to "fire" position and rocker pivot to "automatic" or "full auto" position. Pull trigger and allow bolt to go forward gradually by retarding actuator with left hand. Place gun upside down on knee or table, muzzle to the left. With thumb of right hand depress frame latch and with left hand slide frame toward rear of gun about 1 inch by pushing or tapping on rear grip. Take gun from table or knees, turn it right side up, grasp receiver firmly with left hand near feedway. With right hand grasp grip in normal firing position, index finger on trigger. Pull the trigger as in firing and withdraw frame to rear of gun, pulling down on frame as well as to the rear as it is withdrawn. Failure to pull down as frame is withdrawn causes frame latch to strike against the rear portion of the trigger and trip, damaging these parts.

d. Recoil spring.—Support muzzle of barrel on table or knees with open side of receiver facing operator. Grasp receiver with left hand with thumb in position to depress buffer pilot. With thumb of left hand press down on buffer pilot. With thumb and forefinger of right hand engage flange of buffer and pull down until buffer pilot clears back of receiver. If breech oiler follows, push it back. Hold buffer with thumb and forefinger of right hand and withdraw this entire unit from receiver. Care should be taken to obtain a firm hold on the buffer to prevent the recoil spring, which is compressed, from forcing it out of the operator's hand. When the recoil

6

spring, at rest, measures less than 10½ inches it should be replaced.

e. Bolt, lock, and actuator.—Grasp receiver, bottom up, with left hand. Slide bolt to the rear and lift out. Slide actuator with lock forward and lift out lock. Slide actuator to rear, turn receiver over, and allow actuator to fall out into right hand.

f. Breech oiler.—Hold receiver upside down in left hand, muzzle to front. Hook end of forefinger of right hand under front end of right side of breech oiler. Pull up and to left until right side of breech oiler is out of groove in receiver. Hold the oiler in this position with the right thumb. Without moving thumb, hook forefinger of right hand under front end of left side of oiler and pull oiler out of groove in left side of receiver. With forefinger of right hand pull entire front end of oiler up and to the rear. Lift the oiler out of the receiver, front end first. It is not necessary to remove oiler for ordinary cleaning purposes.

■ 7. DISASSEMBLING OF GROUPS.—*a. Magazine, 50-round* (fig. 3).—Remove winding key and lift off cover. Lift up on end of rotor retainer and slide it out of its engagement in the hub. The rotor assembly may then be removed from body. Further disassembly of magazine is not necessary and is prohibited.

b. Magazine, 20-round (fig. 3).—Slide out floor plate. Hold fingers over bottom of magazine tube to keep magazine spring from flying out. Remove follower.

c. Butt stock (fig. 1).—Complete disassembly of butt stock is not necessary for ordinary cleaning. However, when necessary for replacement of broken or worn parts, butt stock slide and butt plate may be removed by taking out the screws holding them in place.

d. Frame group (figs. 1 and 2).—(1) Hold frame in left hand and, using back end of actuator (not knob) as a tool in right hand, depress short finger of pivot plate and push out rocker pivot with thumb of left hand; lift out rocker and pivot.

(2) Again using actuator but steadying hand with thumb against frame to prevent excessive movement, depress long finger of pivot plate and withdraw safety.

(3) Hold frame upright with grip in right hand. Press simultaneously with both thumbs on sear and trigger pivots. These pivots project sufficiently so that by a quick pressure thereon pivot plate will protrude on the other side far enough to permit withdrawal. While withdrawing pivot plate with left hand, press down on trigger and sear with thumb of right hand to release pressure of springs on pivots. Do not cant pivot plate during withdrawal. The remaining components of the firing mechanism are then free to be removed. Disconnector can be removed from trigger by simply withdrawing it.

(4) To remove magazine catch, rotate it in a counterclockwise direction to its full limit and pull out to the left. Removal should be limited to replacement of broken parts. Removal of magazine catch submits magazine catch spring to unnecessary strain and is apt to damage it.

e. Bolt group (figs. 1 and 2).—(1) Push hammer pin out of bolt from left side; hammer, firing pin, and firing pin spring will then tend to spring out under the impulse of the firing pin spring. Caution should be exercised to prevent these parts from springing away and becoming lost. The firing pin spring must not be stretched.

(2) Extractor should not be removed for ordinary cleaning or disassembling. To do so submits it to unnecessary strain and is apt to cause it to break or become set.

(3) To remove extractor from bolt insert corner of actuator flange under head of extractor on face of bolt and pull extractor out and up to withdraw it from its groove. When disassembling extractor from or assembling it to bolt, do not lift extractor higher than necessary for lug to clear anchorage hold as otherwise setting or breakage may occur.

f. Receiver group (figs. 1 and 2).—For ordinary training, receiver and parts assembled thereto need not be disassembled.

(1) To remove rear sight leaf, drive out sight base pin and remove leaf. Take care to see that rear sight leaf plunger and spring do not fly out.

(2) The ejector can be removed by lifting leaf sufficiently to disengage detent and unscrewing same from receiver. Do not try to unscrew ejector with bolt assembled and in forward position. To do so may damage the ejector and bolt.

(3) Fore grip can be removed by unscrewing fore grip screw.

(4) Barrel should be removed only for purpose of replacement and then only by authorized ordnance personnel.

■ 8. ASSEMBLING OF GROUPS.—In general, groups and their components are assembled and replaced in the gun in the reverse order of that in which they were removed or disassembled. Certain precautions in assembling (*a* and *b* below) must be observed in order that the parts will function properly after gun is assembled.

a. In assembling trigger mechanism, first see that magazine catch is in place. Assemble springs in their proper recesses. Assemble disconnector to trigger by depressing disconnector spring and sliding disconnector into place.

(1) Place trigger, trip, sear, and sear lever in their respective positions in frame, making sure forward end of sear lever rests on tip of disconnector. To aline these parts, hold frame in left hand and press downward with end of thumb on trigger and base of thumb on sear. Insert pivot plate. To avoid binding, apply gentle pressure with ball of right hand over entire pivot plate.

(2) Insert safety as far as it will go, and using actuator as a tool depress long finger of pivot plate and push safety home. Turn safety to "fire" position.

(3) Place rocker in position in frame with flat side against sear lever. Insert rocker pivot as far as it will go. Using actuator as a tool, depress short finger of pivot plate and push rocker pivot home. Turn rocker pivot to "full auto" position. If rocker is assembled backward, the gun will fire full automatic but not semiautomatic.

b. If extractor has been removed, slide it into place, lifting head only enough to clear stud; avoid excessive pressure. Insert firing pin and spring in bolt being careful to avoid stretching firing pin spring. Place hammer in position with rounded edge upward and push hammer pin into place.

■ 9. REPLACING GROUPS IN GUN.—Before replacing groups in gun, be sure ejector is screwed all the way home and that breech oiler is in place.

a. Place receiver on table or knees, botttom up, and insert actuator in receiver, knob to front. Slide actuator forward and place lock in guideways of receiver, with the word "up" correctly readable from the rear and the arrow pointing toward the muzzle. Slide actuator to the rear and place bolt in position by lowering it into the receiver front end first, moving it to the rear, and allowing it to drop into place.

b. Slide bolt forward and start recoil spring, assembled to buffer guide, into its recess in bolt. Push recoil spring down into bolt until buffer pilot clears end of receiver. Let buffer pilot find its seat in receiver and snap into place.

c. For guns equipped with drilled buffer guides, assemble about two-thirds of recoil spring on buffer guide, place retaining pin (nail or strong wire) through coil of spring and hold of buffer guide to hold compressed spring on buffer guide. Care should be taken to see that retaining pin is entered in the hole in buffer guide from the side opposite to that of the milled or flat portion of the buffer flange. Slide the bolt forward and start the loose end of recoil spring into recess in actuator which rides in the bolt. Feed loose end of spring into its recess until rear end of pilot can be seated in hole in rear end of receiver, care being taken to see that buffer disk is in place before pilot is placed in the hole in receiver. Move bolt rearward until rear end of bolt is brought up against spring retaining pin. Remove retaining pin which permits spring completely to seat itself within the recess. Release bolt.

d. Before fitting frame to receiver, be sure that safety is set at "fire" position and rocker pivot at "full auto." Slide frame onto receiver. Frame latch will lock from in position. Holding trigger depressed, operate actuator back and forth several times to test mechanism.

SECTION III

CARE AND CLEANING

■ 10. GENERAL.—The attention given to a weapon of this type determines largely whether it will shoot accurately and function properly. The bore and chamber must be kept in perfect condition for accurate shooting. Also, it is just as important that the receiver and moving parts be kept clean, lubricated, and in perfect condition for efficient functioning. Care and cleaning will include the magazines, which must be kept free from rust, grit, gum, etc., in order to function properly. When not on the person, the submachine gun should be habitually transported in a suitable boot provided with the necessary brackets for attachment.

■ 11. CLEANING AND LUBRICATING.—*a. General.*—Keep the gun well cleaned and oiled. After each day's firing, clean the bore, chamber, and all parts and surfaces of the receiver, bolt, ejector, and extractor that have been in contact with powder gases. Remove frame from receiver and take bolt out; thoroughly clean front end of bolt and extractor. With the bolt removed, the bolt well, the throat of the receiver, and the ejector head are readily accessible.

b. Bore.—(1) As the barrel of the submachine gun is not removed for cleaning, it must be cleaned from the muzzle if the submachine gun cleaning rod is used. However, by using the rifle cleaning rod the barrel can be cleaned from the breech. Push rifle cleaning rod through buffer pilot hole in back of receiver and thread a patch through eye of rod.

CAUTION: In cleaning bore, care must be taken not to foul cleaning patch in slots of recoil compensator.

(2) Run several wet patches through bore. For this purpose water must be used; warm water is good, and warm soapy water is better. Remove patch, attach cleaning brush, and run brush back and forth through bore several times. Care should be used to insure that brush goes all the way through bore before direction is reversed. Remove brush and run several wet patches through bore. Follow this by dry patches until patches come out clean and dry, then saturate a patch in sperm oil and push it through bore.

NOTE—Sperm oil (U.S.A. Spec. 2–45A) should be used when available. When not available, motor oil, weight 20, or any light grade machine oil may be used in an emergency.

c. Chamber.—The chamber should be cleaned with the chamber cleaning brush at reasonable intervals in extended firing to facilitate extraction of cartridge cases and to prevent pitting and rusting. It is not necessary to disassemble gun for this purpose. The brush is introduced through ejection opening in receiver and should be used vigorously. Upon completion of firing, the brush having been used, the chamber is further cleaned and oiled in the process of cleaning the bore.

d. Exterior surfaces.—Wipe off exterior surfaces of gun with a dry cloth to remove dampness, dirt, and perspiration, then wipe off all metal surfaces with an oiled (sperm oil) rag. The stock and grips should be wiped with raw linseed oil.

e. Magazines.—It is imperative that magazines be given the best of care and kept in perfect condition. They should be disassembled, wiped clean and dry, and thinly coated with oil. Dirt that gets into them through careless handling during range or other firing must be removed. Care must be exercised in handling magazines to avoid denting or bending, especially the lips of the mouth of the box type magazine.

f. Lubricating the gun.—To function efficiently, the gun must be properly lubricated. For this purpose use aircraft machine gun lubricating oil (U.S.A. Spec. 2-27) or sperm oil (U.S.A. Spec. 2-45A).

(1) Having removed frame from receiver, oil should be dropped over pivot points of trigger and trip, sear and sear lever, and disconnector and rocker.

(2) Holding receiver in left hand, open side up, bolt should be slightly drawn back and oil dropped on locking lugs of lock, on sides of lock, and on all sliding surfaces of bolt and receiver.

(3) *Felt pads in breech oiler should be kept well-saturated with oil.*

(4) After assembling frame to receiver, bolt should be drawn back and a little oil should be dropped on rounded front end of bolt. Actuator knob should be worked back and forth several times to insure penetration of oil to all parts of mechanism.

(5) *All sliding surfaces should be oiled frequently and freely to insure perfect functioning of the gun.*

SECTION IV

FUNCTIONING

■ 12. GENERAL.—The pressure of the gases generated in the barrel by the explosion of the powder in the cartridge is exerted in a forward direction against the bullet, driving it through the bore, and in a rearward direction against the face of the bolt. This force drives the bolt and the actuator together to the rear against the pressure of the recoil spring. During rearward movement, the processes of unlocking, extracting, ejecting empty shell, and compressing the recoil spring are effected; during the forward movement, the processes of feeding, locking, and firing the cartridge are accomplished. To simplify explanation of functioning, the cycle has been divided into three phases as set forth below.

■ 13. BACKWARD MOVEMENT OF RECOILING PARTS (first phase).—The cartridge having been fired, the pressure from the exploding cartridge is transmitted through the forward end of the bolt to the lock and through the lock to the locking surfaces of the receiver. The powder used is fast burning, so that the highest chamber pressure obtained is nearly instantaneous. The lock being made of bronze and the bolt and receiver being made of steel, the high chamber pressure causes the lock to adhere to the locking surface on the receiver, thus locking the bolt in its forward position until this pressure subsides. As soon as the high chamber pressure has subsided, the lock moves upward, clears the locking surfaces in the receiver, and the bolt can move to the rear. The angle of the lock is such that the moment the lock is moved to clear the receiver locking surfaces there is only sufficient powder pressure in the chamber to force the cartridge case and bolt to the rear, eject the empty case, and compress the recoil spring, which thus stores up energy for the forward movement. The empty case is unseated by the chamber pressure as the bolt is unlocked. As soon as the bolt moves back from the abutment on the under side of the receiver, the firing pin spring forces the firing pin to the rear away from the face of the bolt. The empty cartridge is held on the face of the bolt by the extractor. After the bolt has traveled to the rear about 2 inches the ejector, which protrudes in a groove on the left side of the bolt, comes in contact with the base of the empty cartridge and throws it to the right through the ejection opening. The bolt still has about 1¾ inches to go to the rear before the back of the bolt comes in contact with the buffer. The rearward movement of the bolt, carrying the actuator and compressing the recoil spring, expends nearly all the energy imparted by the chamber pressure, so that the bolt does not strike heavily against the buffer. The buffer pad absorbs the remaining shock. On the under side of the bolt there are two sear notches so that, if the bolt strikes the buffer pad, the rear sear notch will pass over the sear and allow the sear to engage in the front notch. If the movement is not strong enough to cause the bolt to strike the buffer pad, the sear will engage in the rear notch. If the bolt moves to the rear far enough to eject the empty cartridge case and to feed the next cartridge from the top of the magazine, the bolt will normally be back far enough to engage the sear with the rear notch.

■ 14. FORWARD MOVEMENT OF RECOILING PARTS (second phase).—When the trigger is pulled, the bolt moves forward under the action of the recoil spring, carrying the lock and actuator with it. After the bolt moves forward about 1 inch, the forward end of the bolt comes in contact with the back of a cartridge and pushes it forward until the nose of the bullet comes in contact with the bullet ramp in front of the receiver. The lips of the magazine hold the cartridge in a straight line until the cartridge has almost cleared the magazine. The cartridge is guided into the chamber by the bullet ramp and the lips of the magazine. When the cartridge has been seated in the chamber, the extractor snaps around the rim of the cartridge. Just before the bolt reaches its forward position, the lock is cammed down into the locking grooves in the receiver so that the bolt is completely locked as the hammer on the under side of the bolt strikes the receiver. The hammer being of a triangular shape, the lower point strikes the receiver causing the hammer to pivot around the hammer pin and strike the head of the firing pin with the upper point thereby firing the cartridge. The rectangular surface of the bolt, striking the abutment of the receiver, stops the forward movement.

■ 15. ACTION OF TRIGGER MECHANISM (third phase).—*a*. (1) *Rocker pivot set at "single."*—When the trigger is pulled, the trigger rotates around the trigger pivot (the forward pin of the pivot plate) and lifts the disconnector up under the sear lever. The sear lever lifts the front end of the sear; this causes the sear to rotate around the sear pivot (the rear pin of the pivot plate), and in so doing depresses the nose of the sear, disengaging it from the sear notch on the under side of the bolt. As the bolt goes forward, the point of the rocker is in the ᴛ-groove on the under side of the bolt. When the point of the rocker strikes the rear end of the ᴛ-groove, the rocker is forced forward. The rounded part of the rocker comes in contact with the disconnector and forces the disconnector out from under the sear lever. As soon as the disconnector has been disengaged from the sear lever, the sear spring and the sear lever spring force the sear and sear lever up into firing position, so that the sear notch on the bolt will catch on the next rearward movement of the bolt.

(2) *Rocker pivot set at "full auto."*—The rocker pivot is of eccentric design, so that when the rocker pivot is set at "full auto" the rocker is lowered enough to allow the bolt to move forward without striking the point of the rocker. Therefore, the sear remains in its lowered position as long as the trigger is depressed. The rocker pivot cannot be turned from "full auto" to "single" unless bolt is retracted.

b. (1) *Safety set at "fire."*—When the safety is turned toward the front, the flat milled surface is in such a position that the sear is allowed to rotate around the sear pivot.

(2) *Safety set at "safe."*—When the safety is turned toward the rear, the rounded part of the safety engages in a groove in the rear of the sear and locks the sear in its uppermost position. The safety can be turned only when the bolt is to the rear.

c. The magazine catch rotates around its pin and is held down in the engaged position by the magazine catch spring. The stud on the magazine catch is to hold the box type magazine; the drum type is held by the rectangular catch on the left side. The trip functions only when the box type magazine is used. As the magazine empties, a fin on the back of the magazine follower rises up under the trip, causing the trip to rotate around the trigger pin, compressing the disconnector spring, and holding the disconnector forward of the sear lever. Thus the bolt will not go forward on an empty chamber when the box magazine is used.

SECTION V

STOPPAGES AND IMMEDIATE ACTION

■ 16. MISFIRE.—In the event of misfire, retract or cock bolt with a sharp, quick pull on actuator knob. This should insure ejection of misfired cartridge. Inspect chamber to see that it does not contain an unexpended round.

■ 17. OTHER STOPPAGES.—For any other malfunction, retract bolt as above and clear throat and chamber of gun by turning gun over on its side and letting case or cartridge roll out. If necessary, remove magazine and allow cartridge or case to fall out the bottom. While manipulating the gun under these circumstances, always set the gun at "safe."

SECTION VI

SPARE PARTS AND ACCESSORIES

■ 18. SPARE PARTS.—*a.* The parts of any submachine gun will in time become unserviceable through breakage or wear resulting from continuous usage. For this reason spare parts are provided for replacement of the parts most likely to fail, for use in making minor repairs, and in general upkeep of the submachine gun. Sets of spare parts should be maintained as complete as possible at all times and should be kept clean and lightly oiled to prevent rust. Whenever a spare part is used to replace a defective part, the defective part should be repaired or a new one substituted in the spare parts set. Parts that are carried complete should at all times be correctly assembled and ready for immediate insertion in the submachine gun.

b. Twenty- or fifty-round magazines are also issued as spares; the former for use with submachine guns issued for use by cavalry motorcyclists and the latter for submachine guns issued for use by cavalry combat vehicle troops. The quantity of magazines issued per gun is based on the allowance of ammunition authorized. The allowances of spare parts and of magazines are prescribed in SNL A—3L.

■ 19. ACCESSORIES.—*a. General.*—Accessories include the tools required for disassembling and assembling and for the cleaning and preservation of the gun. They must not be used for any purpose other than as prescribed. There are a number of accessories, the names or general characteristics of which indicate their uses or applications. Therefore, detailed description or method of use of such items is not outlined herein. However, accessories embodying special features or having special uses are described in *b* below.

b. (1) *Brush, chamber cleaning, M6.*—The brush consists of a steel wire core with bristles, the core being twisted in a spiral to hold the bristles in place. It is used to clean the chamber of the submachine guns.

(2) *Brush, cleaning, caliber .45, M5.*—The brush consists of a brass wire core with bristles and tip. The core is twisted in a spiral and holds the bronze bristles in place. The brass tip which is threaded for attaching the brush to the cleaning rod is soldered to the end of the core.

(3) *Case, accessory and spare parts, M1918.*—This is a leather, box-shaped case, approximately 2¼ inches wide, 3½ inches high, and 5½ inches long. It is used to carry spare parts and a number of the smaller accessories.

(4) *Rod, cleaning, submachine gun.*—This consists of a long steel rod having a circular loop at one end and a cleaning patch slot at the other end. Permanently affixed to the cleaning patch end of the rod is a head having a threaded hole to receive the cleaning brush, caliber .45, M5.

(5) *Sling, gun, M1923 (webbing).*—The gun sling is fastened to the swivels provided on the gun. It consists of a long and short strap, either of which may be lengthened or shortened as desired to suit the particular soldier using it.

(6) *Thong.*—The thong consists of a tip with cleaning patch slot and a weight tied to the ends of a 30-inch length of cord. It is used in cleaning the bore of the submachine gun.

SECTION VII

INDIVIDUAL SAFETY PRECAUTIONS

■ 20. GENERAL RULES.—*a.* Before firing—

(1) Test trigger mechanism as "safe" and "single."

(2) See that bore is clear and clean.

(3) Work bolt back and forth rapidly several times to see that it is clean, well-oiled, and works freely.

(4) Examine magazines and eliminate faulty ones.

(5) See that each magazine is free from dirt and that it is properly loaded.

b. For range practice, insert loaded magazine only on order of the officer in charge of firing. Do not attach loaded magazine until ready to fire.

c. Carry gun with bolt retracted and safety on "safe" until ordered to attach magazine, and keep safety on "safe" until gun is raised to fire.

d. Keep trigger finger outside trigger guard until gun is raised to fire.

e. From time magazine is attached until gun is cleared and clearance checked, keep gun pointed toward target, whether firing dismounted or from a vehicle.

f. For semiautomatic fire, make certain that rocker pivot is set on "single" and the safety on "safe" before attaching magazine.

g. For full automatic fire, make certain that rocker pivot is set on "full auto" and the safety on "safe" before attaching magazine.

h. For vehicular firing at moving ground targets, the gunner must keep the safety on "safe" at all times when his vehicle is moving.

i. Habitually set the safety at "safe" while changing magazines and during lulls in firing.

j. To clear gun, first remove the magazine.

k. At CEASE FIRING and upon halting at the finish of a vehicular run, set the safety at "safe," remove magazine, and see that no cartridge remains in the chamber before turning away from the firing point.

l. After gun has been cleared and checked for clearance, close the bolt on an empty chamber.

SECTION VIII

AMMUNITION

■ 21. GENERAL.—The information in this section pertaining to the several types of cartridges authorized for use in the Thompson submachine gun, caliber .45, M1928T1, includes a description of the cartridges, means of identification, care, use, and ballistic data.

■ 22. CLASSIFICATION.—Based upon use, the principal classifications of ammunition for this rifle are—

a. Ball, for use against personnel and light matériel targets.

b. Tracer, for observation of fire and incendiary purposes.

c. Dummy, for training (cartridges are inert).

■ 23. LOT NUMBER.—When ammunition is manufactured an ammunition lot number, which becomes an essential part of the marking, is assigned in accordance with specifications. This lot number is marked on all packing containers and the identification card inclosed in each packing box. It is required for all purposes of record, including grading and use, reports on condition, functioning, and accidents in which the ammunition might be involved. Since it is impracticable to mark the ammunition lot number on each individual cartridge, every effort should be made to maintain the ammunition lot number with the cartridges once the cartridges are removed from their original packing. Cartridges which have been removed from the original packing, and on which the ammunition lot number has been lost, are placed in grade 3. It is therefore necessary when cartridges are removed from original packings that they be so marked that the ammunition lot number is preserved.

■ 24. GRADE.—Current grades of existing lots of small arms ammunition are established by the Chief of Ordnance and are published in Ordnance Field Service Bulletin 3-5. No lot other than that of current grade appropriate for the weapon will be fired. *Grade 3 ammunition is unserviceable and will not be fired.*

■ 25. IDENTIFICATION.—*a. Markings.*—The contents of original boxes are readily identified by the markings on the box. Similar markings on the carton label identify the contents of each carton.

b. Color bands.—Color bands painted on the sides and ends of the packing boxes further identify the various types of ammunition. The following color bands for cartridges are used:

Ball........................... Red.
Tracer........................ Green on yellow.
Dummy........................ Green.

c. Types and models.—(1) The cartridges authorized for use in this weapon are designated—

Ball, caliber .45, M1911.
Tracer, caliber .45, M1.
Dummy, caliber .45, M1921.

(2) When removed from their original packing containers, the cartridges may be identified except as to ammunition lot

number and grade by physical characteristics described below:

(*a*) *Ball.*—The bullet of the ball cartridge has a gilding metal jacket.

(*b*) *Tracer.*—The bullet of the tracer cartridge has a gilding metal jacket which is painted red for a distance of approximately $\frac{3}{16}$ inch from the tip.

(*c*) *Dummy.*—There are two designs of dummy cartridges, both of which have tinned cartridge cases. In the one containing an inert primer, a ⅛-inch hole is drilled into the body of the cartridge case. In the other the primer is omitted leaving an opening in the head of the cartridge case thereby obviating the necessity for the ⅛-inch hole in the body. The bullet in both instances is the same as that used in the ball cartridge.

■ 26. CARE, HANDLING, AND PRESERVATION.—*a.* Small arms ammunition is not dangerous to handle. Care, however, must be exercised to keep the boxes from becoming broken or damaged. All broken boxes must be immediately repaired and all original markings transferred to the new parts of the box. The metal liner should be air tested and sealed if equipment for this work is available.

b. Ammunition boxes should not be opened until the ammunition is required for use. Ammunition removed from the airtight container, particularly in damp climates, is apt to corrode thereby causing the ammunition to become unserviceable.

c. The ammunition should be protected from mud, sand, dirt, and water. If it gets wet or dirty wipe it off at once. Light corrosion, if it forms on cartridges, should be wiped off. However, cartridges should not be polished to make them look better or brighter.

d. No caliber .45 ammunition will be fired until it has been identified by ammunition lot number and grade.

e. Do not allow the ammunition to be exposed to the direct rays of the sun for any length of time. This is liable seriously to affect its firing qualities.

■ 27. STORAGE.—Whenever practicable, small arms ammunition should be stored under cover. Should it be necessary to leave small arms ammunition in the open, it should be raised on dunnage at least 6 inches from the ground and the pile covered with a double thickness of paulin. Suitable trenches should be dug to prevent water from flowing under the pile.

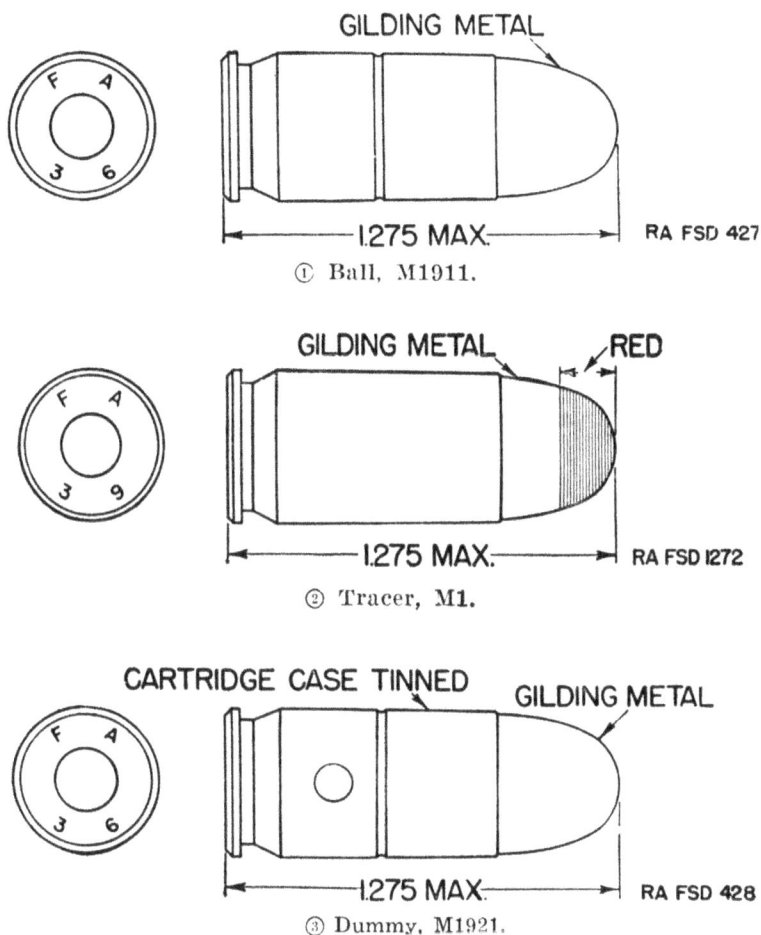

GILDING METAL

1.275 MAX.   RA FSD 427

① Ball, M1911.

GILDING METAL   RED

1.275 MAX.   RA FSD 1272

② Tracer, M1.

CARTRIDGE CASE TINNED   GILDING METAL

1.275 MAX.   RA FSD 428

③ Dummy, M1921.

FIGURE 4.—Cartridges, caliber .45.

■ 28. BALLISTIC DATA.—Approximate maximum ranges are as follows:

Cartridge, ball, caliber .45, M1911 . . . . 1,600 *Yards*
Cartridge, tracer, caliber .45, M1 (estimate only) 1,600

CHAPTER 2

MANUAL OF SUBMACHINE GUN

Section I

MANUAL OF ARMS

■ 29. General.—In general, the manual of the submachine gun is prescribed so as to provide uniform, simple, safe, and quick methods for handling the gun. Precision and simultaneous execution is seldom required; however, a simple and effective manual which can be executed in cadence is included for those occasions when its use is desirable.

■ 30. Carrying Position.—Except when otherwise prescribed, the submachine gun is habitually carried slung over the right shoulder, butt down, barrel to the rear, right hand grasping the sling, hand in front of armpit (fig. 5). For

Figure 5.—Carrying position.

formal drills and ceremonies the box magazine is habitually used. For dismounted marches and for field exercises, the submachine gun may be carried slung over either shoulder. When troops are at ease, the submachine gun is kept slung

22

unless otherwise ordered. When troops are at rest, the submachine gun may be unslung and held in any desired position. In executing ATTENTION the carrying position is assumed. PARADE REST and the RIGHT-HAND SALUTE are executed in the normal manner by releasing the grasp of the right hand from the sling.

■ 31. POSITION OF RAISE ARMS (fig. 6).—Executed at the command: 1. RAISE, 2. ARMS, or when other members of the command are executing the manual of the pistol at the

FIGURE 6.—Position of RAISE ARMS.

command RAISE PISTOL. This position is assumed by the gunner from the position of attention by grasping the fore grip with the left hand, withdrawing the right arm from between the gun and the sling, then assuming the position as shown in figure 6. The gunner standing at attention grasps the rear grip with his right hand, forefinger extended along the outside of the trigger guard; holds the gun with the butt plate resting on his belt over his right hip, barrel extending upward to the front in a vertical plane and at an angle of 45°. He steadies the gun by pressing the stock against his right side with his right forearm. Left arm and hand are at the left side in a natural unconstrained position.

■ 32. Position of Port Arms (fig. 7).—Executed at the command: 1. PORT, 2. ARMS. The gunner standing at attention holds the piece in a vertical plane parallel to and about 4 inches in front of his body, barrel extending upward to the left at an angle of 45°. The right hand grasps the small of the stock. The left hand holds the foregrip and is opposite to and at the same level as the point of the left shoulder.

FIGURE 7.—Position of PORT ARMS.

■ 33. Position of Inspection Arms (fig. 8).—Executed at command: 1. INSPECTION, 2. ARMS, or when other members of the command are executing the manual of the pistol, at the command INSPECTION PISTOL. The position is the same as PORT except that the actuator has been pulled to the rear, opening the belt, and the safety has been set at "safe."

■ 34. Position of Present Arms (fig. 9).—Executed at the command: 1. PRESENT, 2. ARMS. The gunner, standing at attention and grasping the small of the stock lightly with the right hand and the fore grip with the left hand, holds the piece 4 inches in front of the center of his body in such manner that the barrel is vertical and to the rear with the muzzle up. The gunner's right arm should be straight without constraint.

FIGURE 8.—Position of INSPECTION ARMS.

FIGURE 9.—Position of PRESENT ARMS.

■ 35. EXECUTION OF MANUAL.—In describing the execution of the following movements, it is assumed that neither the box nor the drum magazine has been attached to the gun. If desired, the manual can be executed while either magazine is attached, with minor and obvious modifications.

a. RAISE ARMS to PORT ARMS (in two counts).—(1) Carry the gun diagonally across the front of the body so that the barrel and the muzzle are up; at the same time grasp the fore grip smartly with the left hand.

(2) Move the right hand from the rear grip to the small of the stock.

b. PORT ARMS to RAISE ARMS (in three counts).—(1) Remove the right hand from the small of the stock and grasp the rear grip, with the forefinger extending along the outside of the trigger guard.

(2) With the left hand still grasping the foregrip, carry the piece to the position of RAISE.

(3) Drop the left hand smartly to the side.

c. RAISE ARMS to INSPECTION ARMS (in three counts).—(1) Raise the left hand and pull the actuator to the rear, cocking the piece. Set the safety at "safe" using the thumb of the right hand.

(2) Execute the first movement of RAISE ARMS to PORT ARMS.

(3) Execute the second movement of RAISE ARMS to PORT ARMS.

d. INSPECTION ARMS to RAISE ARMS (in four counts).—(1) Execute the first movement of PORT ARMS to RAISE ARMS.

(2) Execute the second movement of PORT ARMS to RAISE ARMS.

(3) With the right thumb move the safety to the fire position, and with the right forefinger pull back on the trigger; the left hand moves to the actuator, grasps it, and allows the bolt to move forward without shock.

(4) Execute the third movement of PORT ARMS to RAISE ARMS.

e. RAISE ARMS to PRESENT ARMS (in three counts).—(1) Execute the first movement of RAISE ARMS to PORT ARMS.

(2) Execute the second movement of RAISE ARMS to PORT ARMS.

(3) Carry the piece smartly to the position of PRESENT ARMS.

f. PRESENT ARMS to RAISE ARMS (in four counts).—(1) Carry the piece smartly to the position of PORT ARMS.

(2) Execute the first movement of PORT ARMS to RAISE ARMS.

(3) Execute the second movement of PORT ARMS to RAISE ARMS.

(4) Execute the third movement of PORT ARMS to RAISE ARMS.

SECTION II

LOADING AND FIRING

■ 36. LOADING MAGAZINES.—*a. Box magazines.*—(1) The normal capacity of a box magazine is 20 cartridges. The cartridges feed into the magazine easily. If for any reason excessive force is required to feed the cartridges out of the magazine, the energy of the bolt is taxed to such an extent that a misfire may result. The forward edge of the magazine is rounded to prevent loading cartridges backward.

(2) The lips of the mouth of the box magazine should be a distance of .55 inch apart. If by accident the magazine mouth should become deformed, the lips should be carefully bent back to this dimension.

b. Drum magazines.—To load a drum magazine, remove the winding key by lifting the flat spring thereon and sliding the key off. The cover can then be removed. Place the cartridges, bullet up, in the spiral track of the body, beginning with a full section at the mouth. The simplest method to begin loading is to fill one outer section and then rotate the rotor until this section reaches the mouth. Thereafter, continue to fill successive sections until the end of the spiral track has been reached. Fill each section completely; do not skip any section and do not fill beyond the end of the spiral track. After the magazine is properly filled, replace the cover and key, and wind to the number of clicks indicated on the magazine name plate. Drum magazines, when wound to the number of clicks indicated on the case, should not be rewound after shots have been fired, as the resultant strong spring tension interferes with the surety of action of the gun, as well as incurring the possibility of breaking the main spring of the magazine.

NOTE.—To fire the dismounted course requires 25 rounds (see par. 54). A drum may be loaded with 25 rounds as follows: proceed as in *b* above except that only the first five sections are filled beginning at the mouth. Replace the cover and key and wind six clicks. This special loading is done under the supervision of an officer or experienced noncommissioned officer, and is prohibited except in preparation for marksmanship firing.

■ 37. LOADING WEAPON.—At the command INSERT LOADED MAGAZINE, the bolt is pulled to the rear position, the safety is turned to safe, and the magazine of the type desired placed in the grooves of the gun. The box type magazine is pushed upward into the groove at the end of the trigger guard until it clicks into place (fig. 10). The 50-round drum magazine is inserted from left to right in the horizontal grooves of the piece (fig. 11). The magazine should always be pushed well up or in so that the magazine catch can snap into position

FIGURE 10.—Inserting box magazine.

and hold the magazine securely. In order to fire, the safety is turned to "fire." When the trigger is pulled, the forward movement of the bolt feeds a round of ammunition from the magazine into the chamber and fires the gun.

■ 38. FIRING.—a. In single shot (semiautomatic) fire, release the trigger after each shot to reengage the sear, as each pull of the trigger releases but a single shot when the fire control is set for "single."

b. In full automatic fire, release the trigger after each burst of two or three shots and realine sights on the target before firing again. Short bursts can be more accurately placed than long bursts.

c. In firing from the box (20-round) magazine, the bolt is automatically held in the open position when the magazine has been emptied. To close the bolt on an empty chamber, remove magazine and let the bolt go forward slowly, retarding the actuator with the left hand. Do not snap the bolt on an empty chamber; ease it gradually into the forward position.

d. In firing the drum (50-round) magazine, the bolt automatically closes on the empty chamber when the magazine has been emptied.

FIGURE 11.—Inserting drum magazine.

e. This gun fires on the forward stroke of the bolt. Unless the magazine is removed before the bolt is released by pulling the trigger, the gun will continue to load itself as long as there is ammunition in the magazine.

f. Correct firing positions for the submachine gun differ from those prescribed for the rifle in that, with the submachine gun, the gunner's right shoulder is shoved strongly forward against the butt of the weapon, and his body is so placed that he faces more directly toward his target then does the rifleman. It is particularly important for quick and accurate shooting with the submachine gun to assume the correct position whenever full automatic fire is used.

CHAPTER 3

MARKSMANSHIP, KNOWN DISTANCE TARGETS

SECTION I

PREPARATORY TRAINING

■ 39. PURPOSE.—The purpose of preparatory training in submachine gun marksmanship is to teach the gunner the essentials of good shooting and to develop fixed and correct shooting habits before he undertakes range practice.

■ 40. FUNDAMENTALS.—To become a good submachine gunner a soldier must be thoroughly trained in the following essentials of good shooting:
Correct sighting and aiming.
Correct range estimation.
Correct positions.

■ 41. PHASES OF TRAINING.—*a.* Marksmanship training is divided into the following phases:
(1) Nomenclature, functioning, and care of the weapon.
(2) Preparatory marksmanship training.
(3) Range practice.

b. The submachine gunner will be made proficient in mechanical training before he receives instruction in marksmanship training.

c. A thorough course in preparatory training precedes any range practice. This preparatory training is given to all soldiers expected to fire the submachine gun during range practice, including those previously qualified.

d. All units armed with the submachine gun will include instruction in mechanical training and fundamental elements of submachine gun marksmanship—sighting, aiming, range estimation, and positions—for all recruits. Instruction commences with the initial instruction of the recruit, in conjunction with other arms and weapons, and continues throughout the period of recruit training.

■ 42. METHODS OF TRAINING.—To insure proper instruction, well-qualified officers and noncommissioned officers are placed in charge of instruction. Instruction must be given by subject in the proper sequence so that the student will employ what he has been taught when he moves to the next step or phase of training. Brief talks followed by demonstrations by the instructors precede any work by the students. No student is progressed from one stage of training to the next stage unless he is qualified. Constant supervision by noncommissioned and commissioned officers, with individual instruction where and when necessary, is essential. The coach and pupil system of instruction is used whenever a man is in the firing position. Instruction and exercises are rotated in order to keep up interest and enthusiasm in the class.

■ 43. EQUIPMENT.—a. The equipment required for preparatory marksmanship training is simple and readily improvised from materials at hand. The following list includes the necessary equipment:

Submachine gun.
Sighting disk.
Submachine gun rest.
Material for blackening sights.
Pencil.
Paper.
Targets E, F, and M.
Cleaning and preserving materials.

b. The equipment required for range practice includes, in addition to the necessary items listed above, the materials necessary to construct and maintain target ranges (see sec. IV, ch. 3, and ch. 5), targets, ranges, weapons, vehicles, and range safety equipment.

■ 44. USE OF SIGHTS.—a. General.—(1) Sights are set by raising the leaf and sliding the slide to the range desired. Lateral correction is obtained by turning the small thumb screw.

(2) The personal element of holding the submachine gun affects the accuracy of fire, especially when on automatic. Consequently there may be considerable variation in hits when the same sight setting is used by different persons. Holding the gun tightly tends toward shooting high.

b. *Windage and drift.*—One point of windage will change the point of strike 1 foot at 100 yards. At ranges of 300 yards or more, lateral correction is also made for drift. The drift table showing drift to right from line of bore with 230-grain bullet is as follows:

Range in yards..	50	100	150	200	250	300	350	400	450	500
Drift in feet......	0	0.10	0.25	0.50	0.85	1.40	2.30	3.25	4.50	5.90

NOTE.—Drift at ranges of less than 300 yards is less than 1 foot and therefore negligible.

c. *Range.*—(1) The weapon is primarily intended for firing at short ranges where quick shooting is required. Therefore the battle sight is normally used. For firing with the battle sight, the sight leaf is laid down and the open sight used. No windage or drift correction is necessary.

(2) The trajectory of the submachine gun bullet is less flat than that of a rifle bullet. It is important that the gunner be taught to allow for this characteristic in aiming and to understand the effect of the trajectory on his line of sight. The following table gives the height of the trajectory at points along the trajectory when the sight is set at various ranges:

HEIGHT OF TRAJECTORY ABOVE LINE OF SIGHT FOR
STANDARD AMMUNITION
[Height of trajectory at points indicated]

Range (yards)	50		100		150		200		250	
	Feet	*Inches*	*Feet*	*Inches*	*Feet*	*Inches*	*Feet*	*Inches*	*Feet*	*Inches*
100	0	$4\frac{13}{16}$		0						
200	1	$4\frac{13}{16}$	1	$10\frac{13}{16}$	1	$6\frac{5}{8}$		0		
300	2	$6\frac{5}{8}$	4	3	5	0	4	$7\frac{3}{16}$	2	$11\frac{3}{8}$
400	3	$11\frac{3}{4}$	7	0	9	$2\frac{3}{8}$	10	$1\frac{7}{16}$	9	$9\frac{5}{8}$
500	5	7	10	$2\frac{3}{8}$	14	0	16	6	17	$8\frac{3}{8}$

Range (yards)	300		350		400		450		500	
	Feet	*Inches*	*Feet*	*Inches*	*Feet*	*Inches*	*Feet*	*Inches*	*Feet*	*Inches*
100										
200										
300		0								
400	8	$2\frac{3}{8}$	4	$10\frac{13}{16}$		0				
500	17	$7\frac{3}{16}$	15	$10\frac{5}{16}$	12	6	7	$1\frac{13}{16}$		0

32

■ 45. SIGHTING AND AIMING.—*a. General.*—In all preparatory exercises involving sighting and aiming and in all
range firing, both sights of the rifle are blackened. The
blackening is done by holding each sight for a few seconds
in the point of a small flame which is of such a nature that
a uniform coating of lamp black will be deposited on the
metal. Materials most commonly used for this purpose are
carbide lamp, kerosene lamp, candles, shoe paste, and stove
polish.

b. Exercises.—(1) *No. 1.*—*(a) Purpose.*—To show the correct alining of sights.

(b) Method.—The instructor places the submachine gun
in the submachine gun rest and alines the rest with a blank
sheet of paper. The instructor then alines the sighting disk
with the sights of the submachine gun by directions to the
marker who controls the disk. When the disk is correctly
alined the instructor commands: HOLD. The instructor
moves away from the submachine gun and directs the pupil
to look through the sights in order to observe the correct aim.

(2) *No. 2.*—*(a)—Purpose.*—To show the importance of
uniform and correct aiming.

(b) Method.—The submachine gun with the sights blackened is placed in the rest and pointed at a sheet of paper
mounted on a box about 50 feet away. The pupil takes
position behind the weapon and looks through the sights
without touching the weapon or the rest. The pupil directs
the marker with the small disk to move the disk until the
sights are correctly alined with the bottom of the disk and
then commands: MARK. The marker without moving the
disk makes a small dot on the paper through a hole in the
center of the disk. The marker then moves the disk to
change the alinement. The same process is repeated until
three dots are made. These dots outline the shot group.
The instructor should discuss with the student the size and
shape of the shot group, pointing out the errors.

■ 46. RANGE ESTIMATION.—*a. General.*—(1) The submachine gunner must be well trained in hasty range estimation
and its application to markmanship. Because the weapon is
normally employed quickly and at short ranges, the follow-

ing methods of range estimation are used:

Estimation by eye.

Observation of fire.

(2) The usual method of range estimation is by eye. The submachine gunner is taught to estimate accurately and fix permanently in his mind two distances, 50 yards and 100 yards. Targets at other ranges are estimated in comparison with these units of measure.

(3) When the effect of a shot or bursts of shots can be seen by the gunner, he corrects the range setting applied by estimation in order to increase the accuracy of his fire.

b. Exercises.—The following exercises can be used as guides in instructing the submachine gunner in range estimation. Ranges used are short and at no time greater than 500 yards. The exercises are especially suitable for class exercises.

(1) *No. 1.—(a) Purpose.*—To familiarize the gunner with the units of measure, 50 yards and 100 yards.

(b) Method.—The units of measure, 50 yards and 100 yards, are staked out on the ground up to 500 yards. The gunner is required to become familiar with the appearance of the unit of measure from the prone, kneeling, and standing positions on the ground and from a moving vehicle in his normal riding position.

(2) *No. 2.—(a) Purpose.*—To give practice in range estimation.

(b) Method.—From a suitable point, ranges are previously measured to normal targets within 500 yards. The gunner is required to estimate the ranges to the various objects as they are pointed out by the instructor and record his estimation on a sheet of paper. At least one-half of the estimates are made from the kneeling and sitting positions. Thirty seconds is allowed for each estimate. When all the ranges have been estimated, the paper is checked by the instructor and the true ranges given to the student.

■ 47. POSITIONS.—*a. General.*—Assuming a correct position in firing, the submachine gun has direct bearing on the effectiveness of fire. The positions from which the submachine gun may be fired are standing, kneeling, prone, and from the hip while marching. The latter is relatively ineffective and should rarely be used. Each gunner is given sufficient practice to enable him to assume all positions rapidly and efficiently.

b. Standing (fig. 12).—The standing position is the position normally used. The left foot is well advanced with the body leaning slightly forward, about two-thirds of its weight support by the left foot. The right foot is firmly planted on the ground in the approximate walking position. The

FIGURE 12.—Standing position.

arms support the piece and are allowed to assume a natural, unconstrained position as shown in figure 12. The cheek is placed firmly against the stock. The trunk of the body is twisted so as to shove the right shoulder forward strongly against the butt of the piece in the direction of fire. The right shoulder must be tense and pushed into the gun by turning the trunk at the hips and humping the shoulder. The recoil is very slight for any one shot, but the rapid accumulation of the successive slight recoils in full automatic fire tends to push the gunner's shoulder backward, if he is not well braced for the thrust, and the muzzle of his gun will move upward and to the right, thus affecting the accuracy of his fire. If the proper position is taken, the gun will not climb and need not be held down with the left hand.

c. Kneeling (fig. 13). The kneeling position affords a

steadier aim than does the standing position and is useful whenever the gunner can crouch behind a rock, log, or other protection. In the kneeling position, the left toe points at the target, the lower leg being practically vertical. The right knee is on the ground, pointing about 45° to the right of the target. In general, this position is similar to the kneeling position prescribed for a rifleman, except that the submachine gunner is facing more directly toward his target

FIGURE 13.—Kneeling position.

so as to force his right shoulder forward against the butt of his gun. The gunner supports most of the weight of the gun with his left hand; it is not necessary to grasp the piece tightly with the left hand. The gunner's right arm assumes a natural position as in the standing position.

d. *Prone* (fig. 14).—The prone position is the steadiest one; it should be used whenever time and terrain permit, particularly for firing at ranges over 100 yards. This position is similar to the prone position prescribed for a rifleman, except that the submachine gunner is facing more directly toward his target so as to force his right shoulder forward against the butt of his gun.

■ 48. MARKSMANSHIP EXERCISES.—*a. First.*—(1) *Purpose.* —This exercise is designed as a preliminary step in teaching the gunner to pick up his target by any slight movement in his general field of vision, and to move the gun rapidly from a standing position to the target with sights properly alined. The gunner simulates firing a shot or two at each of several targets which are exposed in rapid succession. In combat he

FIGURE 14.—Prone position.

is usually required to fire on several point targets in rapid succession; he must aim and fire quickly and accurately.

(2) *Method.*—Place the gunner in the standing position at A (fig. 15), gun cocked, unloaded, and set for semiautomatic fire; targets (sec. IV) are partially concealed and all turned with edges toward the gunner. The gunner is instructed to look generally over his field of fire by shifting his eyes slightly, without focusing his eyes on any particular object.

until he sees one of the targets move. The target is exposed on the order of an instructor who gives the gunner no advance indication of which target is to be used. The gunner brings his line of aim on this target at the proper height and simulates firing one shot. Without moving the gun from his shoulder, the gunner looks generally over his field of fire until he sees another target move and then simulates firing a shot at that target. Targets initially remain exposed about 5 seconds; this is later reduced to 3 seconds. In a short time, a gunner will be able to pick up the movement of a target and simulate fire on it with facility. When this has been accomplished, the gunner moves forward about 10 yards, sets the gun for full automatic fire, and simulates firing short bursts of 2 to 4 shots on each target of groups C and D. These targets initially remain exposed for 10 to 15 seconds; this later is reduced to 5 seconds.

b. Second.—After the exercise described in *a* above has been mastered, the gunner goes through the dismounted practice course as prescribed in section II, using no ammunition. This is repeated until he can go through the entire course with facility in the allotted time.

c. Third.—After the gunner has become proficient in the two exercises described above, he is then practiced in the vehicular courses (figs. 19 and 20) without ammunition. This exercise may be postponed until after he has actually fired the dismounted practice course. In any event, he repeatedly practices the vehicular courses before being permitted to fire them with ball ammunition.

■ 49. EXAMINATION.—*a.* Prior to the date of actually firing the submachine gun for record, each gunner so firing is required to pass the examination of preparatory training shown on the form in paragraph 102. The date of this examination is recorded on each individual's qualification record card.

b. The questions given herein are examples for the examination. Men are required to explain them in their own words or demonstrate them by their own actions. These questions are given only as a guide. Any pertinent question under the subjects listed on the form in paragraph 102 should be asked.

Q. Name the parts of the weapon as I point to them.—*A.* Pupil names each part as pointed out.

Q. How many rounds of ammunition can be placed in the magazines?—*A.* The box magazine holds 20 rounds; the drum magazine holds 50 rounds.

Q. Remove the frame group and show me the extractor.— *A.* Pupil removes frame group and shows the instructor the extractor.

Q. What is the first thing to do to assemble the trigger mechanism?—*A.* See that the magazine catch is in the assembled position.

Q. When should the weapon be loaded, and what should be done before loading?—A. Only when you are ready to fire.   The safety lever should be set to "safe" before loading.

Q. Should the bolt be closed with a loaded magazine inserted?   Why?—*A.* No; because the forward movement of the bolt loads a round into the chamber and fires the weapon.

Q. In case of misfire what should first be done?—*A.* The actuator knob should be pulled to the rear to insure ejection of a misfired cartridge, and the safety should then be turned on.

Q. Demonstrate the standing, kneeling, or prone position. —*A.* Pupil demonstrates the required position.

Q. Where is the battle sight?—*A.* Pupil points out the battle sight.

Q. How do you set a range setting of 200 yards and a right windage of 1 point?—*A.* Pupil demonstrates setting.

Q. Demonstrate how to load a drum magazine.—*A.* Pupil demonstrates method of loading drum magazine.

Q. What parts of the gun should be cleaned after firing each day?—*A.* The bore and chamber, and all parts and surfaces of the receiver, bolt, ejector, and extractor that are contacted by the powder gases

SECTION II

COURSES TO BE FIRED

■ **50. ARMY REGULATIONS APPLICABLE.**—AR 775–10 prescribes details as to who will fire and ammunition allowances.

■ 51 INSTRUCTION PRACTICE.—*a*. The following table prescribes the firing in instruction practice in the order followed by the individual soldier. The table is fired three times for instruction. Additional firing may be done if ammunition allowances permit.

b. There is a 5-second interval after the completion of phase A, at which time the timer blows a whistle and the firer starts walking from position A to position B. Phase B starts 10 seconds after the whistle.

QUICK FIRE—TARGETS E, F, AND M

Phase	Type of fire	Position	Range	Time	Shots
A......	Single shot........	Standing......	15–35 yards..	Each target exposed 3 seconds.	10 shots, 2 per target.
B......	Automatic bursts of 3.	Standing or kneeling.	25–30 yards..	Each group exposed 5 seconds.	15 shots, 3 per target.

■ 52. RECORD PRACTICE.—The table used for instruction practice is fired once for record.

SECTION III

CONDUCT OF RANGE PRACTICE

■ 53. GENERAL.—Organization commanders are responsible for conducting the range practice of their organizations in accordance with the provisions of this manual and applicable Army Regulations. All firing is done under the direct supervision of a commissioned officer. No person is permitted to start range practice until after he has successfully passed the gunner's examination as shown in preparatory training.

■ 54. PROCEDURE. (fig. 15).— *a. Preparation.*—The officer in charge of firing, the timer, and five target operators take position in rear of point A (fig. 15). The assistant timer and two target operators take position in the pit at point B (fig. 15). The gunner takes his place at the initial position (point A, fig. 15), facing toward the pit, with a drum magazine loaded with 25 rounds of ammunition (see note, par. 36). The officer in charge of firing commands: LOAD. At this command the gunner pulls the bolt to the rear position, turns the safety to "safe," inserts the loaded drum magazine, sets his piece for semiautomatic fire, turns the safety to "fire," assumes the position of "raise arms," and calls "Ready."

b. Phases.—After the gunner calls ready, the officer in charge of firing blows his whistle to initiate phase A. At this signal the gunner assumes the standing firing position.

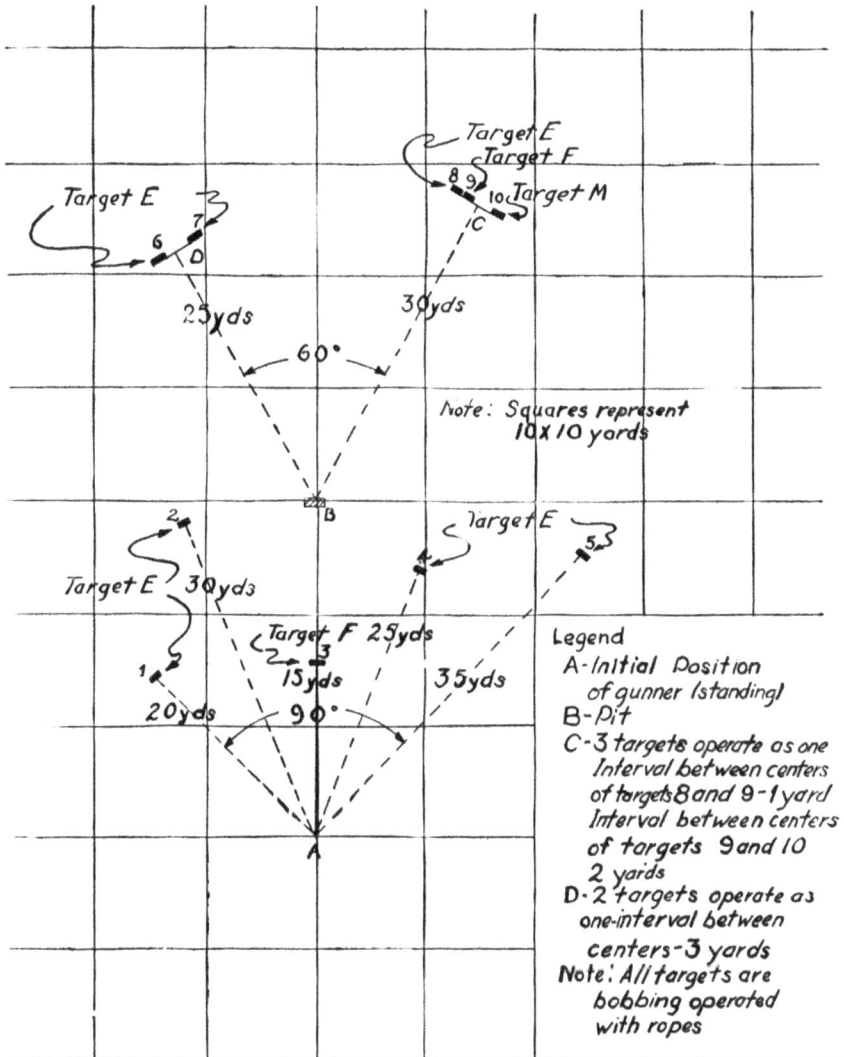

FIGURE 15.—Dismounted practice course.

(1) *Phase A.*—(*a*) Three seconds after the starting whistle the targets in group A (targets 1 to 5, inclusive) are exposed in irregular order, one at a time, for 3 seconds each. Three seconds after the first target disappears, another ap-

pears, etc. The targets are controlled from in rear of the gunner as explained in section IV. The officer in charge of firing designates the target to be exposed by showing a number of fingers equal to the number of the target, for example, three fingers for target No. 3. The timer informs the target operator by hand signal when to expose the target and when to withdraw it. The gunner is allowed two shots per target in this phase, both of which must be fired during a single exposure of the target. The standing position is used. The gunner may keep the gun at his shoulder at all times.

(b) After the fifth target disappears the gunner sets his gun for full automatic fire. Five seconds after the fifth target disappears the timer blows his whistle to initiate phase B.

(2) *Phase B.*—(a) When the timer's whistle sounds, the gunner starts walking toward the pit (B). Ten seconds after the whistle signal, target group C (or D) is exposed for 5 seconds, during which time the gunner fires a burst of three shots at each target of the group. Five seconds after the target group C (or D) disappears, the remaining group D (or C) is exposed for 5 seconds, during which time the gunner fires his remaining ammunition.

(b) Between exposures of group C (D) and group D (C) the gunner may remain halted or he may continue to walk forward, but he may not advance beyond a barrier just short of pit (B).

(c) In phase B the gunner may fire standing or kneeling.

(d) The targets are controlled from the pit (B) as explained in section IV. An assistant timer located in the pit selects the order of exposure of groups C and D and is responsible that the time intervals after the timer's whistle sounds are as prescribed in b(1) above.

c. Action on completion of course.—Upon completion of the course the gunner removes the drum magazine, unloads, and reports that his gun is clear. This fact is verified by an officer or noncommissioned officer before anyone is permitted to move in front of the gun.

d. Timer.—(1) An officer or noncommissioned officer times each run. He is responsible for signaling for the exposure and withdrawal of each target in group A at the proper time intervals and for the second whistle signal prescribed in *b*(1) above.

(2) An assistant timer in the pit (B, fig. 15) is responsible for the exposure and withdrawal of target groups C and D according to the time intervals prescribed in *b*(2) above.

e. Coaches.—(1) *Instruction practice.*—Considerable time and effort can be saved if well-qualified coaches are used at firing points in instruction practice to instruct the gunners. As all fire is timed and as the run is not interrupted once it is started, best results are obtained by having the coach observe the actions of the gunner during each run and then, after the targets are marked, having both gunner and coach report to a designated place where the coach points out errors and gives the necessary instruction to prevent their repetition on the next run. Prior to, during, and after the run, coaches require gunners to employ habitually all practical safety measures

(2) *Record practice.*—No coaching is permitted during record practice.

f. Possible score.—100 points.

g. Form for score card.—See paragraph 102.

h. Classification.—The individual classification to be attained and the minimum aggregate scores required for qualification are as prescribed in AR 775–10.

■ 55. RANGE FIRING.—*a. Marking, scoring. etc.*—(1) The targets are designated 1, 2, 3, 4, and 5 from left to right in group A; 6 and 7 in group D; and 8, 9, and 10 in group C (see fig. 15). The number of men detailed as markers is left to the discretion of the officer conducting the firing; normally they will be the men who pull the targets in phase A. The markers take position in rear of position A (fig. 15) and await the completion of phase B. After phase B is completed, they run to their designated targets, examine them, and face the scorer. One noncommissioned officer detailed as scorer waits at a convenient place. The markers then call out in numerical order the hits or misses; for example, No. 1, a hit or 2 hits; No. 2, a miss, etc. When so instructed by the officer in charge of firing, the markers cover any shot holes with pasters. Two markers then run

43

to groups D and C, respectively, mark the shots, call, and paste them.

(2) During record practice scores are recorded by a non-commissioned officer of a different organization, if practicable, under the supervision of the officer in charge of firing, who personally checks the hits and misses on the targets, checks the score card, authenticates it, and retains all cards in his personal possession except while the man is actually firing.

b. Value of hits.—Each target hit counts 5 points, and each hit on any target (not to exceed two for phase A and three for phase B) counts 2 points.

c. Defective cartridges and malfunctions.—If a defective cartridge or malfunction causes an interruption of the fire, the firer assumes the position of *Raise arms* and indicates this fact to the officer in charge of firing, who commands: CEASE FIRING, and time is stopped. The malfunction is reduced. If the stoppage occurs in phase A before the first shot is fired on any particular target, the full time of 3 seconds is allowed. If the stoppage occurs just before the second shot is fired, 2 seconds are allowed. If the stoppage occurs in phase B, the unconsumed portion of the time allowed for that target group plus 1 additional second are allowed. When ready to resume firing after a stoppage, the firer informs the officer in charge of firing and assumes the appropriate firing position with targets *not* exposed. The officer in charge of firing signals the timer to expose the target and timing is resumed from the instant when the target or group of targets is completely exposed.

SECTION IV

TARGETS, RANGES, AND RANGE PRECAUTIONS

■ 56. TARGETS (fig. 16).—*a.* Target F is a drab silhouette representing a prone figure.

b. Target E is a drab silhouette representing a kneeling figure.

c. Target M is a drab silhouette representing a standing figure. It is in two parts, the upper is target E and the lower is a trapezoidal piece whose upper edge is placed closely against the lower edge of target E.

■ 57. Construction of Range.—*a. Location.*—The course may be laid out on any ordinary terrain, preferably with some grass, weeds, and low underbrush. Targets may be partially concealed or may be near possible concealment in order to represent as nearly as practicable actual enemy groups, but they must be readily discernible to the gunner.

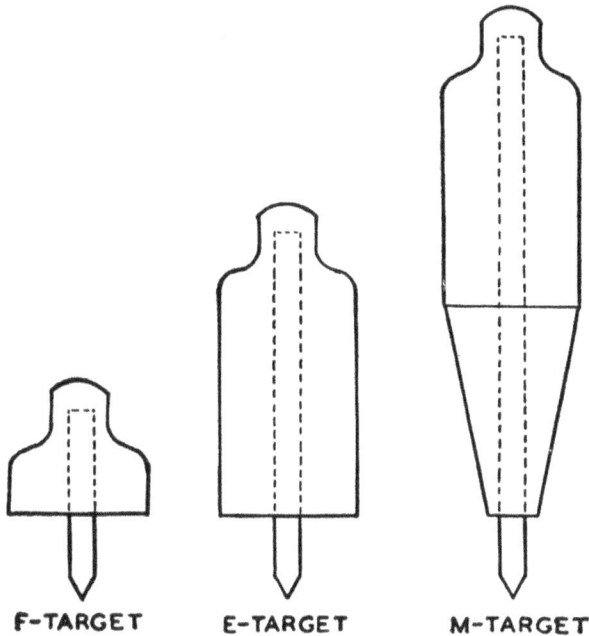

F-TARGET   E-TARGET   M-TARGET

Figure 16.—Targets.

b. Lay-out.—(1) The course is laid out as shown in figure 15. Distances and targets are as indicated in the figure. There are two groups of targets, each necessary for one phase of the firing. Group A is a group of five targets, 4 kneeling (E) and 1 prone (F), which are engaged from position A. Group B consists of two groups (D and C, fig. 15)—one to the left front consisting of two kneeling targets (E); and one to the right front consisting of one standing target (M), one kneeling (E), and one prone (F). In group B, the whole group (C or D) operates as a unit in appearing and disappearing.

(2) The pit consists of an emplacement of concrete (or other suitable material) proof against .45 caliber ammunition, with suitable rear and overhead cover but with an open-

45

ing in front. It is about 3 feet by 6 feet, with the floor at least 3 feet in the ground and the roof not over 1 foot above the level of the ground. It should hold two men and the necessary equipment to control the targets of groups C and D.

(3) If surveying instruments such as transit or aiming circle are available, the angles from A and B (fig. 15) may be readily laid out. These angles may be laid out with a tape measure as follows:

(*a*) Drive a stake at point A, 30 yards from the pit.

(*b*) Along line AB drive a stake 10 yards from A. From this stake along a line at right angles to AB lay off distances of 30 feet and 12½ feet in both directions and drive four stakes. Lay off from point A on lines through these five stakes the distances to the targets as shown in figure 15.

(*c*) Drive a stake 10 yards from point B in prolongation of line AB. From this stake along a line at right angles to line AB lay off distances of 17½ feet in both directions and drive stakes. Lay off from point B on lines through these two stakes the distances to point C and D as shown in figure 15.

c. Operation.—(1) *General.*—All targets are bobbing targets, that is, they are so arranged that they can be fully exposed to the firer, or turned so that only the edge of the target points toward the firer during the time that the target is not exposed.

(2) (*a*) *Group A.*—Each target in this group must be capable of being operated individually from in rear of the firing point. This can be done with two ropes to a target, or by one rope only if a spring arrangement is used to hold the target with only the edge exposed. The use of one rope, which is run through a pipe, to each target will prevent the gunner from knowing which target is to be exposed next. If visible ropes are used, they must be kept tight during the time the gunner is at the firing point.

(*b*) *Group B.*—In this group, the targets at D and at C each operate as a unit. For example, when the group at D is operated, both the E targets are exposed and withdrawn together. The targets are operated from the pit (B, fig. 15). Other details are as in (*a*) above.

(3) *Method of signaling.*—As the officer in charge, the timer, and the target operators are in rear of the gunner (in phase A), the officer can signal the target he desires to be exposed next by the number of fingers he exposes. (See par. 55 for numbering of targets.)

d. Artificial butts.—If an artificial butt is constructed as a bullet stop it should be of earth not less than 30 feet high with a slope of not less than 45°. It should extend about 10 yards beyond the limits of firing and should be as close to the targets as practicable. The slope should be sodded.

e. Hills as butts.—A natural hill to form an effective butt should have a slope of not less than 45°. If originally more gradual, it should be cut into steps, the face of each step having that slope.

f. Numbering targets.—Each target should be designated by a number.

■ 58. RANGE PRECAUTIONS.—(See also pars. 20 and 38.) During firing, all personnel, including marking details, must be in safe positions. The necessary range guards are posted and danger flags prominently displayed before firing begins. There must be no firing until so ordered by the officer in charge. The provisions of AR 750–10 must be complied with during all range firing.

CHAPTER 4

MARKSMANSHIP, MOVING GROUND AND AIR
TARGETS

SECTION I

MOVING GROUND TARGETS

■ 59. GENERAL.—All units armed with the submachine gun will be trained to fire at moving targets, both vehicular and personnel. Normally such fire will be delivered at short ranges in short bursts of fire. The high rate of fire and ability of the submachine gunner to move the trajectory of fire at will make the submachine gun particularly effective against moving personnel, either individual or in groups. Training of the submachine gunner must be such as to enable him to employ his gun effectively and quickly. To this end he must be trained in the proper use of sights and methods of leading at short ranges.

■ 60. SIGHTS.—Moving targets are seldom exposed for long periods and can be expected to move at maximum speed during periods of exposure. Moving personnel are especially difficult to hit. Since the submachine gun is essentially a short range weapon of opportunity, the battle sight is habitually used in firing at moving targets. No windage adjustment is attempted. When necessary to fire near the maximum effective range (about 350 yards) or at greater range and time is available, the rear sight scale may be set at the proper range.

■ 61. LEADS.—Targets that cross the line of sight require the gunner to aim ahead of the target so that the paths of the target and bullet will meet. The distance ahead of the target is called the "lead." Targets which approach directly toward the gunner or recede directly from the gunner require no lead. For personnel targets moving across the line of sight, the point of aim should be slightly in front of the body and the lead corrected by observation of the effect of the spray of the rounds fired.

■ 62. DETERMINATION OF LEADS.—The lead necessary to hit a moving vehicle is dependent upon the speed of the target, the range to the target, and the direction of movement with respect to the line of sight. Moving at 10 miles an hour a vehicle moves approximately its own length of 5 yards in 1 second. The velocity of a bullet from the submachine gun is approximately 800 feet or slightly more than 250 yards in 1 second. Therefore to hit a vehicle moving at 10 miles an hour at ranges of about 250 yards and 500 yards, the leads should be 5 yards and 10 yards, respectively. At a speed of 20 miles an hour the leads should be 10 yards and 20 yards, respectively, etc.

■ 63. APPLICATION OF LEADS.—*a.* Leads are applied by using the length of the target as it appears to the gunner as the unit of measure. This eliminates the necessity for corrections due to the angle at which the target crosses the line of sight, because the more acute the angle the smaller the target appears and the less lateral speed it attains.

b. The following lead table for vehicles is furnished as a guide:

Miles per hour	Range		
	125 yards or less	250 yards	500 yards
10	½ TL	1 TL	2 TL
20	1 TL	2 TL	4 TL

■ 64. TECHNIQUE OF FIRE.—*a.* The following technique is used by the gunner for firing at moving targets:

(1) *Approaching or receding targets.*—The gunner holds his aim on the center of the target and fires automatic in short bursts.

(2) *Crossing vehicular targets.*—The gunner estimates the proper number of leads, alines his sights on the bottom of the target at its rearward point, swings straight across the target to the estimated lead, and fires short bursts of fire keeping the weapon at the proper lead.

(3) *Crossing personnel targets.*—The gunner takes aim

slightly in front of the center of the body of the target and fires short bursts. The lead is changed an appropriate amount after observation of the effect of the spray of bursts.

b. The high rate of fire of the submachine gun allows the gunner to spray the target with fire and improve his lead estimation by actual observation of the effectiveness of his fire.

■ 65. PLACE IN TRAINING.—Firing at moving targets with service ammunition should follow instruction in known distance firing and firing of the dismounted practice course.

SECTION II

AIR TARGETS

■ 66. GENERAL.—Combat arms take the necessary measures for their own immediate protection against low flying hostile aircraft. All available weapons are normally employed in this defense. Consequently all units armed with the submachine gun will be trained in the use of this weapon against air targets. The low muzzle velocity, short effective range, and the tactical employment of the submachine gun are factors which mitigate against its effective use on air targets. Normally the submachine gun is issued as a supplemental weapon to vehicle crews, and more effective weapons, such as the caliber .30 and caliber .50 machine guns, are available and will be employed for immediate fire at aircraft. In comparison with these other weapons, submachine gun fire is relatively ineffective against hostile aircraft. Proper training in the use of sights and methods of leading will increase the effectiveness of fire and are essential phases of the training of the submachine gunner. Application of the fundamentals of firing against moving ground targets is satisfactory for training the submachine gunner in firing against air targets. Special use of sights and leads for air targets are included in paragraphs 67 to 70 inclusive.

■ 67. SIGHTS.—The battle sight is habitually used for firing on low flying airplanes. Hostile airplanes move so rapidly that time is not available for setting of sights.

■ 68. LEADS.—In order to hit an air target in flight, it is necessary to aim an appropriate distance ahead of it and on its projected path of flight so that the target and the bullet will meet. This distance ahead of the airplane is called "lead." A lead must be applied to all firing, except when the target is at an extremely close range (100 feet), when it is diving directly at the gunner, or flying directly from him.

■ 69. DETERMINATION OF LEADS.—*a.* The lead necessary to engage any target depends upon—

(1) Speed of the target.

(2) Range to the target.

(3) Time of flight of the bullet.

(4) Direction of flight of the target with respect to the line of fire.

b. When a target appears, it is impossible for submachine gunners to consider all the factors listed above and to compute accurately the lead required for firing. Use of a quick rule of thumb and experience and proficiency in firing are essential.

c. Based on an average plane of 30 feet in length moving at 200 miles an hour, the plane moves 10 times its own length in 1 second. The velocity of a bullet from the submachine gun is approximately 800 feet or slightly more than 250 yards in 1 second. Therefore to hit an airplane moving across the front at 200 miles an hour at a height of 800 feet, the aiming point of the gunner is 10 leads in front of the airplane, and at a height of 400 feet the aiming point is 5 leads in front of the airplane.

■ 70. APPLICATION OF LEADS.—a. Since it is impracticable for the submachine gunner to estimate accurately the speeds and heights of attacking airplanes, a general rule is given for firing at air targets. Application of this rule by the individual, coupled with the proper fire distribution of other weapons, will form a mass of fire of which some portion will be effective against the air target.

b. The following lead table for air targets is given as a guide:

Miles per hour	Height of plane		
	100 to 600 feet	600–1,000 feet	Over 1,000 feet
200	5 TL	10 TL	20 TL

c. Leads are applied by using the length of the target as it appears to the gunner as the unit of measure. The ability of the submachine gunner to spray his bursts of automatic fire constitutes a balance against the errors in leading by the above general rule.

■ 71. TECHNIQUE OF FIRE.—The following technique should be used by the gunner in firing at air targets:

a. For direct-diving airplanes, direct-climbing airplanes, or airplanes at less than 100 feet height, the gunner holds his aim on the center of the airplane and fires bursts of automatic fire.

b. For all other airplanes, the gunner aims at the airplane, swings to the proper number of leads along the path of the airplane, and fires short bursts of fire, keeping the weapon at the proper lead.

■ 72. PLACE IN TRAINING.—Firing at air targets with service ammunition is the last phase of instruction for the submachine gunner and follows instruction in vehicular firing described in chapter 5.

SECTION III

MOVING TARGETS, RANGES, AND RANGE PRECAUTIONS

■ 73. GENERAL.—The ability of the submachine gunner to hit a target moving on the ground or in the air is developed through appropriate exercises conducted as part of the combat firing of his organization. The following courses are included as exercises to achieve this result. In the moving ground target course, the gunner fires at a moving ground target from a vehicle that moves between bursts but is halted when the gunner actually fires. In the air target course, the gunner fires from a stationary vehicle. Firing at towed air targets follows this course.

■ 74. TARGETS.—The target consists of a rectangular frame 5 feet by 8 feet in size, longer axis horizontal, covered with target cloth or other light-colored material mounted on a suitable carriage which has the ability to move at a maximum speed of 20 miles per hour. The substitution of E and M targets for the target cloth provides a suitable target for training in firing at moving personnel. The addition of a superstructure at the desired height holding a small air target provides a suitable target for training in firing at air targets.

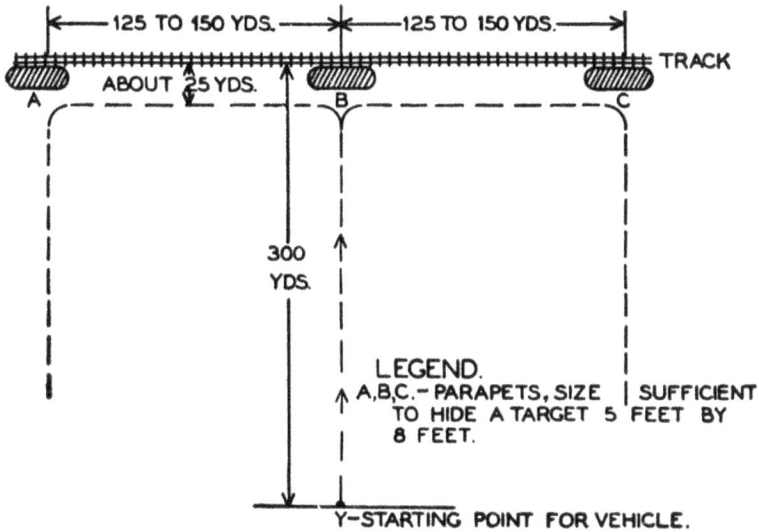

FIGURE 17.—Moving ground target.

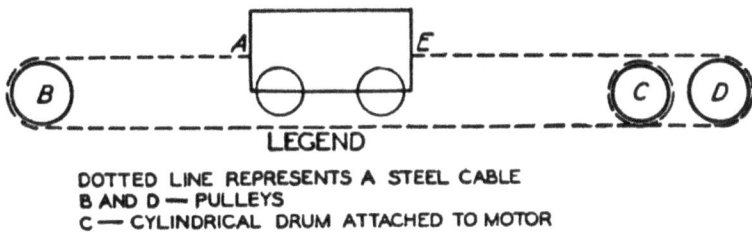

FIGURE 18.—Moving target set-up.

■ 75. RANGE CONSTRUCTION.—*a.* The moving target range should be constructed generally as shown in figures 17 and 18.

b. A, B, and *C* (fig. 17) represent parapets of sufficient size to hide completely the target from the gunner's view, with dugouts in the rear for sheltering pit details. These parapets should be 125 to 150 yards apart. The entire distance from *A* to *C* should be equipped with a suitable continuous track behind the parapets for movement of the target. Y is the starting point of the gunner's vehicle, and YB the direction in which the gunner's vehicle moves. The track for the vehicle should be smooth and level. The entire area from Y to the parapets should be clear in order that ground firing points may be employed for instruction in firing at moving personnel and air targets.

c. The dotted line (fig. 18) represents a steel cable which is fastened to the target carriage. The cable is run from (A) along the top of the track to and around a pulley (B); then back under the target track and carriage to a cylindrical drum (C) around which it goes twice; then around a pulley (D) which is mounted on a movable frame to adjust tension in the cable; then back to the target carriage at (E). The shaft (or axle) of drum (C) is attached through a transmission and clutch to a motor. Thus the drum may be rotated in one direction by running the motor with the transmission in reverse and in the other direction with the transmission in a forward speed. The target carriage is equipped with flanged wheels and is run on narrow gage tracks.

d. When the necessary material is not available for the construction and operation of such an installation, a simpler arrangement may be made by towing a double-ended sled behind a vehicle; but with a sled, the target cannot be so readily stopped and started again in either direction.

■ 76. OPERATION OF RANGE AND COURSE TO BE FIRED FOR MOVING GROUND TARGETS.—a. (1) *First run.*—The gunner starts at Y mounted in the vehicle with his weapon loaded and locked on automatic; the vehicle moves at 15 to 20 miles per hour toward B. The target may be behind either parapet A or C. Upon telephone or visual signal from behind the starting line, the target is released and moved at 10 to 20 miles an hour to parapet B. When the target appears, the gunner's vehicle is stopped, the gunner unlocks his weapon and fires from the stationary vehicle at the moving target as long as it is visible. The target should be released so that the gunner fires the first run at a range of not more than 250 or less than 150 yards.

Rounds fired: 20.

Method of fire: Bursts of about 3 rounds fired in rapid succession.

(2) *Second run.*—The gunner quickly reloads with a new clip of 20 rounds, locks the piece, and calls "Ready," where-

upon the vehicle again moves forward at 10 to 20 miles per hour. The target should be held behind parapet *B* after the first run for 30 seconds, at the end of which time it is again released in either direction. During the second run, the gunner again fires from a stationary vehicle as before as long as the target is visible.

Rounds fired: 20.

Method of fire: Optional.

b. Scoring.—10 points for hitting the target; 2 points for each hit on the target. Maximum score: 100.

■ 77. OPERATION OF RANGE AND COURSE TO BE FIRED FOR AIR TARGETS.—*a.* (1) *First run.*—The gunner stands at a ground firing point immediately in front of parapet *B* with the submachine gun loaded and locked on automatic. The moving target modified with a small superstructure of the desired height is placed behind parapet *B*. Upon signal by the officer in charge of firing, the target is released and moves toward either parapet *A* or *C*. Five seconds after the target has begun to move, the gunner may commence firing and may continue to fire until his ammunition is exhausted.

Rounds fired: 20.

Method of fire: Optional.

(2) *Second run.*—The gunner stands at a ground firing point immediately in front of parapet *B* with the submachine gun loaded and locked on automatic. The moving target, modified with a small superstructure of the desired height, is placed behind parapet *A* or *C*. Upon signal by the officer in charge of firing, the target is released and moves toward parapet *B*. Five seconds after the target has begun to move, the gunner may commence firing and may continue to fire until his ammunition is exhausted.

Rounds fired: 20.

Method of fire: Optional.

b. Scoring.—10 points for hitting the target on each run; 2 points for each hit on the target. Maximum score: 100.

■ 78. RANGE PRECAUTIONS.—(See pars. 20, 38, and 58.) Firing takes place only at such times as a flag is properly displayed by the target detail indicating that they have taken adequate cover. In instruction firing at air targets and firing at towed air targets, the right and left limits of fire will be plainly marked by posts or flags.

CHAPTER 5

VEHICULAR FIRING

SECTION I

GENERAL

■ 79. GENERAL.—*a.* Units armed with the submachine gun are trained to fire from stationary and moving vehicles at appropriate targets. Practice in firing from moving vehicles follows instruction in known distance firing, dismounted. Two instruction courses are described in this chapter, one for firing from open vehicles and one for firing from turreted vehicles. For vehicular firing, vehicles are closed as for combat except that ports may be open. Speeds are as nearly uniform as practicable and generally conform to speeds indicated in each course. The vehicle is equipped with the full number of weapons authorized, all in normal position for combat.

b. The courses outlined in sections II and III, below, are furnished as guides and the direction of fire, the length and contour of the track, and arrangement of and ranges to the targets may be varied to suit local conditions. For instance, the three phases outlined for each type of vehicle may be fired separately. It is desirable that all personnel at a post, camp, or station fire the same courses.

SECTION II

OPEN VEHICLE

■ 80. ARMY REGULATIONS APPLICABLE.—AR 775–10 prescribes details as to who will fire and ammunition allowances.

■ 81. INSTRUCTION PRACTICE.—The following table prescribes the firing in instruction practice in the order followed by the individual soldier. One 50-round drum is allowed to fire the table. The table is fired twice. Additional firing may be done if ammunition allowances permit.

QUICK FIRE—TARGETS E AND F

Phase	Type of fire	Range	Vehicular speed	Shots
A	Automatic	75–25 yards	20 miles per hour	15
B	Automatic	65–5 yards	20 miles per hour	10
C	Semiautomatic	12–75 yards	Halted and 20 miles per hour	25

■ 82. RECORD PRACTICE.—Vehicular firing is not included in firing in record practice.

■ 83. PROCEDURE (fig. 19).—*a. Firing phases.*—(1) The gunner takes his place in the front compartment of the vehicle on the track about 50 yards from point *A* with a loaded 50-round drum.

(*a*) *Phase A.*—At the order of the officer in charge of firing, the gunner loads his piece, sets it for automatic fire, and calls "Ready." The officer in charge directs the vehicle to start on the course. The vehicle gains and maintains a uniform speed of approximately 20 miles per hour between halts. The gunner may start firing on the targets of group 1 as soon as the vehicle passes point *A*, at which point the timer starts taking time for the run. The gunner remains low in the vehicle and exposes himself only enough to fire about fifteen rounds at the targets in group 1. He ceases firing on these targets upon reaching point *B* which is 75 yards from point *A*.

(*b*) *Phase B.*—The vehicle continues on the course, and after making the turn and reaching the point *E* the gunner may commence firing on targets of group 2, all to his right front. He fires about ten rounds at targets of this group. He ceases firing when his vehicle passes the point *G*.

(*c*) *Phase C.*—After passing group 2, the gunner moves to the rear of the vehicle, sets his weapon on semiautomatic, and prepares to engage targets of group 3. The vehicle continues on the course until it reaches point *J* where it halts for 15 seconds. The gunner can commence firing on targets of group 3 as soon as the vehicle comes to a full halt.

He may continue to fire after the vehicle resumes moving on the course, but ceases firing upon expenditure of the 50-round drum or when the vehicles passes point *K*.

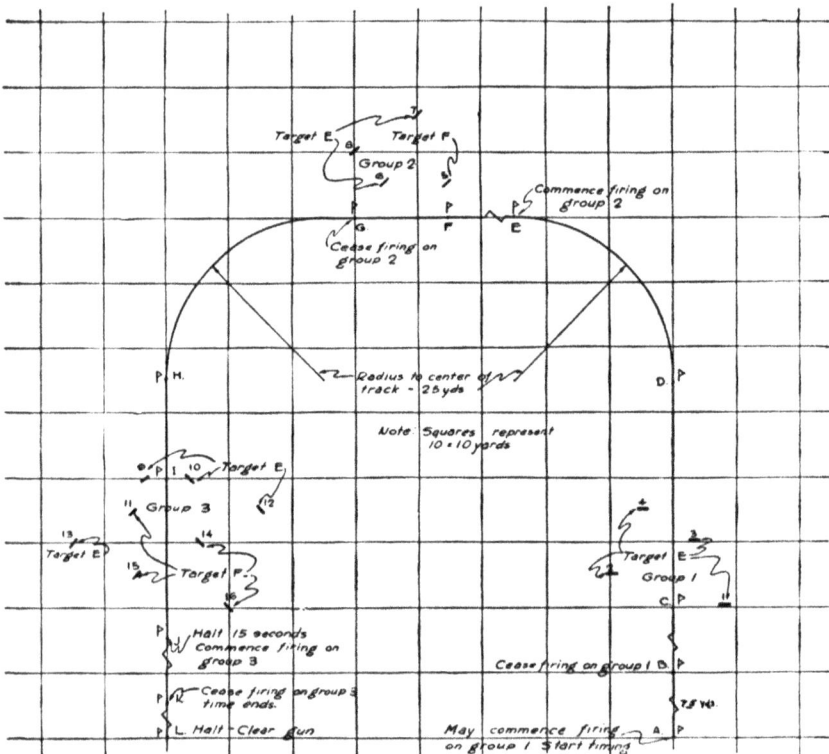

NOTES.—1. All targets are fixed.

2. Targets not shown perpendicular to track are set at an angle of 45° to it.

3. Distances: A to B, 75 yards.   J to K, 50 yards.
               B to C, 25 yards.   K to L, 75 yards.
               E to F, 50 yards.   Others as shown.

FIGURE 19.—Course for firing from open vehicles.

(2) All firing in all phases must cease upon or before the expiration of the time necessary to travel the distance from point *A* to point *K* at 20 miles an hour, plus 25 seconds (65 seconds for the course outlined herein).

b. Action on completion of course.—Upon completion of the course at point *L*, the gunner removes the drum, unloads, and reports that his gun is clear. This fact is verified by a noncommissioned officer before anyone is permitted to move in front of the gun.

c. Timer.—The timer takes position in the vehicle so as

not to interfere with the gunner. He records the time for the entire run from point *A* to point *K* and reports the time for each individual gunner to the officer in charge. The timer may assist the gunner in commencing and stopping his fire at the designated points in the course by tapping the gunner lightly on the back or shoulder or commanding CEASE FIRING at the proper points during the run. He times the halt periods and indicates to the driver when to move.

d. Coaches.—The duties of coaches for vehicular firing are the same as those prescribed for dismounted firing as described in paragraph 54*d*.

e. Marking, scoring, and defective ammunition.—The targets are numbered as indicated in figure 19. The method of marking and scoring and the treatment of defective cartridges and malfunctions are generally the same as prescribed in the dismounted practice course for record and described in paragraph 55.

f. Value of hits.—Each target hit counts 5 points, and each hit on any target (not to exceed 3 hits per target, except on target number 16 which may have 5 hits) counts 2 points.

g. Penalties.—For any infraction of the rules of procedure laid down herein, the gunner is penalized 5 points for each round fired improperly.

h. Possible score.—180 points.

i. Form for score card.—See paragraph 102.

■ 84. CONSTRUCTION OF COURSE.—*a. Safety requirements.*— The extent of the danger area for firing this weapon is 1,600 yards in the direction of fire plus a distance to be determined locally to provide for ricochets. Lateral limits of the safety area must extend 5° beyond the right and left limits of fire. Designation of the lateral limits should take into account the instability of the gunner's mount and the resultant probability of wild shots. The limits of the danger area may be modified by local authorities with the approval of the corps area commander when the nature of the terrain or artificial barriers makes a smaller area safe. (See AR 750–10.)

b. Location. The range may be located on any ordinary terrain with some grass, weeds, and underbrush and a good field of fire for the gunner. Targets may be partially concealed but must be easily located by the gunner. The track should be constructed so as to be reasonably smooth, particularly along those portions where firing occurs.

c. Laying out range.—(1) The area required for the range is about 145 yards by 180 yards. The lay-out should provide about 50 yards additional length to permit the vehicle to acquire the required speed before passing point *A*. The range should be carefully measured according to the distances shown in figure 19. Key points along the track, marked *A* to *L,* inclusive, are marked with small flags. A stake about 12 inches high is driven at the location of each target to insure correct ranges and to facilitate placing the targets.

(2) There are three groups of fixed targets, each necessary for one phase of firing. Group 1 is a group of 4 targets (all targets E, kneeling); group 2 is a group of 4 targets (2 targets E, kneeling, and 2 targets F, prone); group 3 is a group of 8 targets (4 targets E, kneeling, and 4 targets F, prone).

(3) Ranges to the targets are as shown in the following table:

Target No.	Measured from point	Distance along track	Distance from center of track
		Yards	Yards
1	C	0	8
2	C	5	10
3	C	10	3
4	C	15	5
5	F	0	5
6	F	10	5
7	F	5	15
8	F	15	10
9	J	25	3
10	J	25	3
11	J	20	5
12	J	20	15
13	J	15	15
14	J	15	5
15	J	10	5
16	J	5	10

d. Danger signals.—One or more danger signals (red flags) are placed near the range to warn passers-by when firing is in progress. They should be placed on roads or on the crest of hills where they can be seen plainly by those passing.

Section III

TURRETED VEHICLE

■ 85. Army Regulations Applicable.—AR 775–10 prescribes details as to who will fire and ammunition allowances.

■ 86. Instruction Practice.—The following table prescribes the firing in instruction practice in the order followed by the individual soldier. One 50-round drum is allowed for the table. The table is fired twice. Additional firing may be done if ammunition allowances permit.

QUICK FIRE—TARGETS E, F, AND M

Phase	Type of fire	Range	Vehicular speed	Shots
A	Automatic	90–10 yards	20 miles per hour	20
B	Semiautomatic	10–40 yards	Halted and 20 miles per hour	20
C	Semiautomatic	15–70 yards	20 miles per hour	10

■ 87. Record Practice.—Vehicular firing is not included in firing in record practice.

■ 88. Procedure (fig. 20).—a. Firing phases.—(1) The gunner takes his place in the vehicle on the track about 50 yards from point A with a loaded 50-round drum. All submachine gun firing is through pistol ports.

(a) Phase A.—At the order of the officer in charge of firing, the gunner loads his piece, sets it for automatic fire, and calls "Ready." The officer in charge directs the vehicle to start on the course. The vehicle gains and maintains a uniform speed of approximately 20 miles per hour between halts. The gunner may start firing on the targets of group 1 as soon as the vehicle passes point A, at which time the timer starts taking time for the run. The gunner fires about twenty rounds between points A and B, using short bursts of automatic fire on each target. At point B he ceases firing on targets of group 1.

61

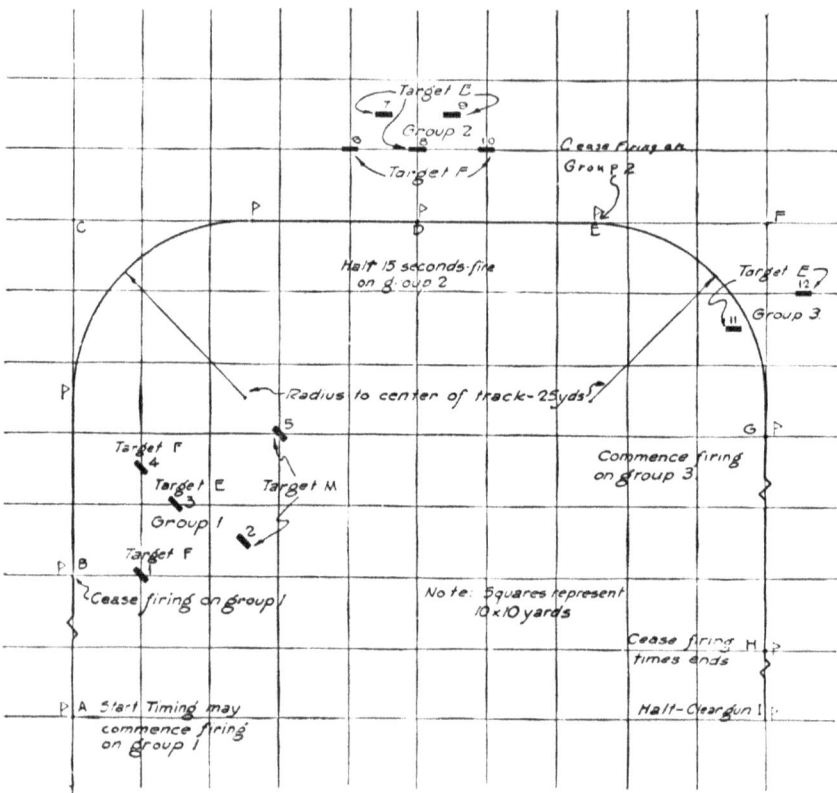

NOTES.—1. All targets are fixed.

2. Targets of group 1 are set at an angle of 45° to the track; those of group 2, parallel to the track; those of group 3, perpendicular to line *FG*.

3. Distances: *A* to *B*, 75 yards.　　*H* to *I*, 45 yards.
　　　　　　　G to *H*, 50 yards.　　Others as shown.

FIGURE 20.—Course for firing from turret-top vehicles.

(*b*) *Phase B.*—As the vehicle moves toward point *D* the gunner sets the gun on semiautomatic and prepares to fire from the left and rear ports. Upon reaching point *D* the vehicle halts for 15 seconds. When the vehicle has come to a full halt, the gunner may commence firing on the targets of group 2. The vehicle moves forward after the 15-second halt. The gunner may continue to fire after the vehicle has moved from point *D*, but must cease firing when point *E* is reached.

(*c*) *Phase C.*—The vehicle continues to move toward point *G*, and the gunner prepares to fire semiautomatic fire from the rear ports on targets of group 3. He may commence firing on these targets when the vehicle passes point G, but must cease firing when the vehicle reaches point H.

(2) All firing in all phases must cease on or before the expiration of the time necessary to travel the distance from point A to point H at 20 miles an hour, plus 25 seconds (57 seconds for the course outlined herein).

b. *Action upon completion of course.*—Upon completion of the course at point I, the gunner removes the drum, unloads and reports that his gun is clear. This fact is verified by a noncommissioned officer before anyone is permitted to move in front of the gun.

c. *Timer.*—The timer takes position in the vehicle so as not to interfere with the gunner. He records the time for the entire run from point A to point H and reports the time for each individual gunner to the officer in charge. The timer may assist the gunner in commencing and stopping his fire at the designated points in the course by tapping the gunner lightly on the back or shoulder or commanding CEASE FIRING at the proper points during the run. He times the halt periods and indicates to the driver when to move.

d. *Coaches.*—The duties of coaches for vehicular firing are the same as those prescribed for dismounted firing as described in paragraph 54d.

e. *Marking, scoring and defective ammunition.*—The targets are numbered as indicated in figure 20. The method of marking and scoring and the treatment of defective cartridges and malfunctions are generally the same as prescribed in the dismounted practice course for record and as described in paragraph 55.

f. *Value of hits.*—Each target hit counts 5 points, and each hit on any target (not to exceed 4 hits per target, except on target number 12 which may have 6 hits) counts 2 points.

g. *Penalties.*—For any infraction of the rules of procedure laid down herein, the gunner is penalized 5 points for each round fired improperly.

h. *Possible score.*—160 points.

i. *Form for score card.*—See paragraph 102.

■ 89. CONSTRUCTION OF COURSE.—a. *Safety requirements.*—The extent of the danger area for firing this weapon is 1,600 yards in the direction of fire plus a distance to be determined locally to provide for ricochets. Lateral limits of the safety area must extend 5° beyond the right and left limits of fire. Designation of the lateral limits should take into account the instability of the gunner's mount and the resultant probability of wild shots. The limits of the danger area may be modified by local authorities with the approval of the corps area commander when the nature of the terrain or artificial barriers makes a small area safe. (See AR 750–10.)

b. Location.—Requirements of terrain for the range are the same as for open-top vehicles (see par. 84*b*). With obvious modifications the same general area may be used for both types of firing.

c. Laying out range.—(1) The area required for the range is about 115 yards by 150 yards. The lay-out should provide about 50 yards additional length to permit the vehicle to acquire the required speed before passing point *A*. The range should be carefully measured according to the distances shown in figure 20. Key points along the track marked *A* to *I*, inclusive, are marked with small flags. A stake about 12 inches long is driven at the location of each target to insure correct ranges and to facilitate placing the targets.

(2) There are three groups of fixed targets, each one necessary for one phase of firing. Group 1 is a group of 5 targets (1 target E, kneeling; 2 target F, prone; and 2 target M, standing); group 2 is a group of 5 targets (3 target E, kneeling, and 2 target F, prone); group 3 is a group of two targets (both target E, kneeling).

(3) Ranges to the targets are as shown in the following table:

Target No.	Measured from point	Distance along track	Distance from center of track
		Yards	Yards
1	B	0	10
2	B	5	25
3	B	10	15
4	B	15	10
5	B	20	30
6	D	10	10
7	D	5	15
8	D	0	10
9	D	5	15
10	D	10	10
11	G	15	5
12	G	20	5

d. *Danger signals.*—One or more danger signals (red flags) are placed near the range to warn passers-by when firing is in progress. They should be placed on roads or on the crest of hills where they can be seen plainly by those passing.

CHAPTER 6

TECHNIQUE OF FIRE

■ 90. CHARACTERISTICS OF FIRE.—*a. General.*—The characteristics of the submachine gun govern the manner in which it is to be used. It is essentially an individual weapon like the rifle, pistol, or bayonet; but it does not supplant or replace any of the others. It is highly effective at close quarters, because it is light, compact, and, with a 50-round drum, can remain in action when other weapons have expended their ammunition. The gun is very dependable due to the simplicity of its mechanism. Efficient gunners can be quickly developed. The gunner has 20 to 50 rounds immediately available, and by changing magazines he can pour a considerable volume of fire into an enemy that confronts him. This volume of fire, and the flexibility of fire of which the individual gunner is capable, are important characteristics of the submachine gun, because of the ease of reloading and shifting the fire when fighting at close quarters. The fact that the weapon can be fired either as a single-shot gun or with full automatic fire is also important. In terrain where cover and concealment are abundant, especially against riflemen or machine gunners, this flexibility of fire is an advantage. The gun is equipped with a compensator which takes up much of the recoil and permits the gunner who uses the weapon properly to attain considerable accuracy of fire even when full automatic fire is used.

b. Comparison with other weapons.—As compared to the M1903 or M1 caliber .30 rifle, the caliber .45 submachine gun is at a disadvantage in that it has less killing power, it is less accurate at ranges over 150 yards, and has no bayonet. However, the submachine gun has the distinct advantage of being well suited to combat in close quarters.

c. Collective firing.—Collective firing is the combined firing of a group of individuals. The submachine gun is normally issued as a vehicular or individual weapon. It is not issued to all members of a unit as is the pistol or the rifle. Consequently, collective firing of submachine guns is not employed. The submachine gun may be used in conjunction with other weapons, especially the machine gun. When used in this manner it is normally employed at short-range targets, while the alternate weapon fires at relatively long range. The weapon may be used by motorcycle scouts operating individually at extremely short ranges.

■ 91. TYPE OF FIRE ORDERS.—*a.* Formal fire orders are seldom necessary or desirable. For control on the target range such orders as INSERT LOADED MAGAZINE, READY, COMMENCE FIRING, and CEASE FIRING are used. In combat, fire orders if necessary at all are normally limited to COMMENCE FIRING, and CEASE FIRING.

b. The target designation may be added to the fire order when the target has not been discovered by the submachine gunner. In this case, the fire order may be given to the submachine gunner as follows, "Jones, rifleman behind tree to right front, commence firing."

■ 92. TARGET DESIGNATION.—The normal employment of the submachine gun is such as to preclude the necessity for the formal target designation desirable for longer-range weapons employed in groups. Targets which are not perfectly obvious may be designated by pointing, by oral description, or by firing upon them with tracer ammunition. Normally targets are discovered and immediately taken under fire by the individual gunner himself.

■ 93. RANGE ESTIMATION.—*a. General.*—In battle, ranges are seldom known in advance, so that the effectiveness of fire depends in large measure upon the accuracy of range estimation. The gunner is trained in the estimation of ranges up to about 500 yards for proper sight setting. He is also made proficient in determining the correct point of aim with the battle sight at all ranges up to about 200 yards.

b. Appearance of objects.—In some cases much of the ground between the observer and the targets is hidden from view. In such cases the range is estimated by the appearance of objects. Whenever the appearance of objects is used as a basis for range estimation, the observer must make allowance for the following effects:

(1) Objects seem nearer—

(*a*) When the object is in a bright light.

(*b*) When the color of the object contrasts sharply with the color of the background.

(c) When looking over water, snow, or a uniform surface like a wheat field.

(d) When looking downward from a height.

(e) In the clear atmosphere of high altitudes.

(f) When looking over a depression most of which is hidden.

(2) Objects seem more distant—

(a) When looking over a depression most of which is visible.

(b) When there is a poor light or fog.

(c) When only a small part of the object can be seen.

(d) When looking from low ground upward toward higher ground.

c. *Training in range estimation.*—Proficiency in range estimation can be obtained only by constant practice and diligence. The methods and exercises prescribed in paragraph 46 are suitable means of developing proficiency in range estimation.

■ 94. SAMPLE PROBLEMS.—a. *General.*—In preparing exercises involving the use of the submachine gun, advantage is taken of field exercises and maneuvers to present logical situations, some phases of which would require the employment of this weapon, both from the ground and from the vehicle. These exercises should include the use of the submachine gun in the dismounted reconnaissance of a road block, employment on outpost duty or in establishing march outpost, and use by motorcycle scouts and by armored vehicle personnel in assumed ambush situations.

b. *Exercises.*—The following exercises are given as a guide and may be modified to suit the terrain, equipment, and time available. Each problem utilizes natural terrain features, equipment normally available, and actual personnel targets. The exercise should be conducted under the supervision of a commissioned officer who will make all checks and point out all errors. Service ammunition is not fired during these exercises. The purposes of all exercises is to train the individual submachine gunner and unit leaders in target designation, range estimation, fire orders, and setting of sights. Every effort is made to carry out the fundamentals of concealment, camouflage, and scouting and patrolling in the conduct of these exercises. Soldiers acting as personnel targets should be rotated with gunners taking the course, and type targets should be shifted frequently to avoid monotony.

(1) *No. 1.*—A stretch of terrain not to exceed 400 yards in length and containing as many natural features as possible, such as trees, shrubs, tall grass, ditches, and walls, is selected for the course. Along a designated path are placed actual personnel targets at various ranges from the path and in normal concealment. Typical targets should include prone, kneeling, and standing soldiers, individual and groups of moving men, machine guns with normal crews, and mounted scouts. The gunner is required to proceed down the designated path and locate targets, estimate ranges, set sights, take position, and simulate fire on each target that he locates. He is accompanied by an instructor who checks all phases of the gunner's action on each type target.

(2) *No. 2.*—In a suitably selected location, a road block of any type should be established and held by a detachment of machine guns and riflemen. Either the submachine gunner dismounted, the motorcyclist with submachine gun, an armored vehicle with submachine gun as an alternate weapon, or any combination of these, should operate against the road block. A commissioned officer should accompany the individual or the vehicle to check and instruct in procedure and criticize the commands of the individual or car commander during the problem.

(3) *No. 3.*—A small area in which buildings predominate and which can be presumed to be a village or city street should be selected for this exercise. The area should be such as to allow personnel to occupy buildings, roofs and windows, and to erect barricades. Personnel armed with the submachine gun and mounted on motorcycles and in armored cars should be required to operate against personnel in buildings and to reduce the barricade. This type of problem is especially beneficial in the training of mounted and dismounted action, collective firing with other weapons, and proper leadership. All actions by individuals, squad or platoon leaders, and individual units should be carefully checked by a commissioned officer, and the exercise should be reviewed and criticized immediately upon completion.

CHAPTER 7
ADVICE TO INSTRUCTORS

■ 95. PURPOSE.—The provisions of this chapter are to be accepted as a guide. They have not the force of regulations. They are particularly applicable to emergency conditions when large bodies of troops are being trained under officers and noncommissioned officers who are not thoroughly familiar with approved training methods.

■ 96. MECHANICAL TRAINING.—*a*. As a general rule instruction is so conducted as to insure the uniform progress of the platoon and company.

b. The instructor briefly explains the subject to be taken up and demonstrates it himself or with a trained assistant.

c. The instructor then causes one man in each squad or subgroup to perform the step while he again explains it.

d. The instructor next causes all members of the squads or subgroups to perform the step, checked by their noncommissioned officers. This is continued until all men are proficient in the particular operation, or until those whose progress is slow have been placed under special instructors.

e. Subsequent steps are taken up in like manner during the instruction period.

■ 97. MARKSMANSHIP.—*a*. *General*.—Training is preferably organized and conducted as outlined in chapters 3, 4, and 5. Officers should generally be considered as the instructors of their units. As only one step is taken up at a time, and as each step begins with a lecture and a demonstration showing exactly what to do, the trainees, although not previously instructed, can carry on the work under the supervision of the instructor.

b. *Instructors*.—It is advantageous to have all officers and as many noncommissioned officers as possible trained in advance in the prescribed methods of instruction. When units are undergoing marksmanship training for the first time, this is not always practicable nor is it absolutely necessary. A good instructor can give a clear idea of how to carry on the work in his lecture and demonstration preceding each step. In the supervision of the work following the demonstration, he can correct any mistaken ideas or misinterpretations.

c. *Equipment.*—The instructor should personally inspect the equipment for the preparatory exercises before the training begins. A set of model equipment should be prepared in advance by the instructor for the information and guidance of the organization about to take up the preparatory work.

d. *Inspection of submachine gun.*—No man is required to fire with an unserviceable or inaccurate weapon. All submachine guns are carefully inspected far enough in advance of the period of training to permit organization commanders to replace all inaccurate or defective weapons.

e. *Ammunition.*—The best ammunition available is reserved for record firing, and the men should have a chance to learn their sight settings with that ammunition before record practice begins. Ammunition of different makes and of different lots should not be used indiscriminately.

f. *Vehicles and drivers.*—The best vehicles and drivers of each organization are made available when vehicular firing begins. Vehicles are thoroughly inspected mechanically and should be suitable for the type of firing desired.

g. *Ranges.*—All ranges to be used, including those for moving targets and air targets, are carefully inspected far enough in advance of the period of use to permit changes or repairs when necessary. Targets and other equipment will be in the best state of repair possible when range practice begins. Arrangements for firing at towed air targets are made with the proper authorities in sufficient time to insure the necessary targets and towing airplanes at the proper time.

■ 98. ORGANIZATION OF WORK IN MARKSMANSHIP.—*a. Preparatory training.*—The field upon which the preparatory work is to be given is selected in advance and a section of it assigned to each organization. The equipment and apparatus for the work will be on the ground and in place before the lecture is given, so that each organization can move to its place and begin work immediately and without confusion.

b. *Range practice.*—(1) The range work is so organized that there is a minimum of lost time on the part of each man. Long periods of inactivity while awaiting a turn on the firing line will be avoided. For this purpose no more men should be on the range at one time than the number of targets or ranges available can accommodate efficiently.

(2) In moving ground and air target firing and vehicular firing, it is advisable to have on the line the order that is next to fire and to have them practice with dummy ammunition or simulated fire. When the size of the firing point makes this action impracticable, each order should be given a score of simulated fire before firing with ball cartridges.

■ 99. LECTURES AND DEMONSTRATIONS IN MARKSMANSHIP.— *a.* The lectures at the beginning of each step are an important part of the instructional methods. The lectures may be given to the assembled command or group.

b. The lecturer will know in advance what he is going to say on the subject. Under no circumstances will he read to a class an outline or lecture prepared by himself. If the instructor cannot talk interestingly and instructively on each subject without the use of copious notes, he should not be giving the lectures at all.

c. The one important thing is to show the men undergoing instruction, by explanation and demonstration, just how to go through the exercises and to impress upon them why they are given.

■ 100. TECHNIQUE OF FIRE; RANGE ESTIMATION.—*a. General.*—The instructor will secure the necessary equipment, lay out and supervise the construction of the courses to be run, and detail and train the necessary assistants prior to the first period of instruction. Instructors use their initiative in arranging additional exercises to those given in paragraph 94. It will be explained to pupils how the exercises used illustrate the principles in the technique of fire. Good work in the conduct of the exercises as well as errors will be called to the attention of all pupils.

b. Range estimation.—The instructor makes the necessary arrangements to set up the training courses for range estimation given in paragraph 93 prior to the first period of this training. All equipment such as paper and pencils will be available to each group of pupils at the beginning of each instruction period.

■ 101. ALLOTMENT OF TIME.—A suggested allotment of time for conducting range practice is tabulated below. The time indicated is for a troop or similar organization rather than for individual firing.

Subject: *Time (hours)*

    Instruction, dismounted practice course...   9

    Instruction, vehicular firing............. 12

    Instruction, moving ground and air targets 12

    Record, dismounted practice course.......   3

■ 102. RECORD OF INSTRUCTION AND QUALIFICATION.—A convenient form for keeping a record of the status of each individual's training is shown below:

RECORD OF SUBMACHINE GUN INSTRUCTION

. . . .
  (Last name)   (First name)   (Middle initial)   (Army serial No.)

Grade. . . . .   Organization. . . . .

Preparatory Training and Gunner's Test

	Date	Result	Initials
General description of weapon			
Removal of groups			
Replacement of groups			
Disassembling of groups			
Assembling of groups			
Care, cleaning, and oiling			
Functioning, including use of safety and fire control			
Stoppages and immediate action			
Spare parts, name and explain use			
Positions, standing, kneeling, prone			
Adjustment and use of sights			
Range estimation, effect of wind and light			
Safety precautions, individual			
Loading and unloading, clip and drum			
Examination			

METHOD OF MARKING (under "Result")

Excellent: ✔ ✔   Unsatisfactory: X
Satisfactory: ✔   Inferior: XX

[Front]

RECORD OF SUBMACHINE GUN INSTRUCTION

. . . .
  (Last name)   (First name)   (Middle initial)   (Army serial No.)

Grade. . . . .   Organization. . . . .

Marksmanship Training and Record Firing

Marksmanship exercises (par. 48)				Date	Initials
First					
Second					
Third					
Preliminary firing	Targets hit	Number of hits	Score		
Dismounted practice					
Dismounted practice					
Dismounted practice					
Vehicular firing					
Vehicular firing					
Moving ground targets					
Air target					
Record firing					Signature
Dismounted practice					

Qualification (dismounted practice course) :
    Expert.
    1st class gunner.
    2d class gunner.   . . . .
                           (Signature of organization commander)

                     . . . .
                               (Grade and organization)
                [Back]

INDEX

INDEX

O

www.ingramcontent.com/pod-product-compliance
Lightning Source LLC
Chambersburg PA
CBHW052201090426
42741CB00010B/2363

9 7 8 1 9 3 7 6 8 4 1 9 8